THE
AMAZING STORY
OF QUANTUM MECHANICS

THE AMAZING STORY OF QUANTUM MECHANICS

A MATH-FREE EXPLORATION OF THE SCIENCE THAT MADE OUR WORLD

JAMES KAKALIOS

GOTHAM BOOKS

GOTHAM BOOKS
Published by Penguin Group (USA) Inc.
375 Hudson Street, New York, New York 10014, U.S.A.
Penguin Group (Canada), 90 Eglinton Avenue East, Suite 700, Toronto, Ontario M4P
2Y3, Canada (a division of Pearson Penguin Canada Inc.) · Penguin Books Ltd, 80 Strand,
London WC2R 0RL, England · Penguin Ireland, 25 St Stephen's Green, Dublin 2, Ireland
(a division of Penguin Books Ltd) · Penguin Group (Australia), 250 Camberwell Road,
Camberwell, Victoria 3124, Australia (a division of Pearson Australia Group Pty Ltd) ·
Penguin Books India Pvt Ltd, 11 Community Centre, Panchsheel Park, New Delhi—
110 017, India · Penguin Group (NZ), 67 Apollo Drive, Rosedale, North Shore 0632, New
Zealand (a division of Pearson New Zealand Ltd) · Penguin Books (South Africa) (Pty) Ltd,
24 Sturdee Avenue, Rosebank, Johannesburg 2196, South Africa

Penguin Books Ltd, Registered Offices: 80 Strand, London WC2R 0RL, England

Published by Gotham Books, a member of Penguin Group (USA) Inc.

First printing, October 2010
10 9 8 7 6 5 4 3 2 1

Copyright © 2010 by James Kakalios
All rights reserved
Pages 315–316 constitute an extension of the copyright page.

Gotham Books and the skyscraper logo are trademarks of Penguin Group (USA) Inc.

LIBRARY OF CONGRESS CATALOGING-IN-PUBLICATION DATA
has been applied for.

ISBN 978-1-592-40479-7

Printed in the United States of America
Set in Trump Mediaeval
Designed by Elke Sigal

*To Thomas, Laura, and David,
who truly make the future*

CONTENTS

SECTION 4: WEIRD SCIENCE STORIES

SECTION 5: MODERN MECHANICS AND INVENTIONS

SECTION 6: THE WORLD OF TOMORROW

Our citizens and our future citizens cannot share properly in shaping the future unless we understand the present, for the raw material of events to come is the knowledge of the present and what we make it.

LIEUTENANT GENERAL LESLIE R. GROVES

(WHO OVERSAW CONSTRUCTION OF THE PENTAGON AND WAS
CHIEF MILITARY LEADER OF THE MANHATTAN PROJECT)

FROM THE FOREWORD TO *Learn How Dagwood Splits the Atom*
WRITTEN BY JOHN DUNNING AND LOUIS HEIL AND DRAWN BY JOE MUSIAL
(KING FEATURES SYNDICATE, 1949)

Quantum Physics? You're Soaking in It!

Perhaps you share my frustration that, well into the twenty-first century, we still await flying cars, jet packs, domed underwater cities, and robot personal assistants. From the 1930s on, science fiction pulp magazines and comic books promised us that by the year 2000 we would be living in a gleaming utopia where the everyday drudgery of menial tasks and the tyranny of gravity would be overcome. Comparing these predictions from more than fifty years ago to the reality of today, one might conclude that, well, we've been lied to.

And yet . . . and yet. In 2010 we are able to communicate with those on the other side of the globe, instantly and wirelessly. We have more computing power in our laptops than in the room-size computers that were envisioned in the science fiction pulps. We can peer inside a person, without the slice of a knife, performing medical diagnoses using magnetic resonance imaging. Touch-activated computer screens, from the local ATM to the iPhone, are everywhere. And the number of automated devices we deal with in a given day is surprisingly high—though none of them look like Robby the Robot.

What did the all those rosy predictions miss? Simply put, they expected a revolution in energy, but what we got was a revolution in information. Implicit in the promise of jet packs and death rays is the availability of lightweight power supplies capable of storing large amounts of energy. But the ability of batteries to act as reservoirs of electrical energy is limited by the chemical and electrical properties of atoms. Scientists and engineers are extremely clever in developing novel energy-storage systems, but ultimately we can't change the nature of the atoms. Information, however, requires only a medium to preserve ideas and intelligence to interpret them.

Moreover, information can endure for thousands of years— consider the long-term data storage accomplished by the Sumerians, whose cuneiform writing on clay tablets enables us to learn about their accounting systems and read the epic tale of Gilgamesh from four thousand years ago. These dried clay tablets, currently held in modern-day Iraq, are fairly bulky, and to share information from them the ancient Sumerians had to transport the actual tablets. But today you don't have to go to Iraq to read the Sumerian tablets—you can view them on the Internet, or someone could send images of them to you instantly via a cell phone camera.

These advances in content storage and transmission were made possible by the development of semiconductor devices, such as the transistor and the diode. Back when the science fiction pulp magazines were first published, data manipulation proceeded via bulky vacuum tubes; the first computers employed thousands of such tubes, along with relay switches consisting of glass tubes filled with liquid mercury. The replacement of these tubes and mercury switches with semiconductor devices enabled an exponential increase in computing power accompanied by a similar decrease in the size of the computer. In 1965 Gordon Moore noted that approximately every two years the number of transistors that could be incorporated onto an integrated circuit doubled. This trend has held up for the past forty years and underlies the technological innovations that define our modern life: from book-size radios in the 1950s to an MP3 player no larger than a stick of gum in 2005; from a cell phone the size of a brick in the 1970s to one smaller than a deck of cards today. These advances in miniaturization have come with continued improvements in the ability to preserve and manipulate information. (If energy storage also obeyed Moore's law, experiencing a doubling in capacity every two years, then a battery that could hold its charge for only a single hour in 1970 would, in 2010, last for more than a century.)

With no transistors, computers would still require bulky vacuum tubes, each one generating a significant amount of heat as it regulated electrical currents. A modest laptop computer currently employs approximately more than a hundred million solid-state transistors for data storage and processing. If all of these transistors were replaced with vacuum tubes, each one a few inches long and at least an inch wide, their physical dimensions, and the need to space them apart to avoid overheating, would yield a vacuum tube computer larger than the White House. Obviously, few institutions aside from the federal government and the largest corporations could afford such a massive computing device. We would consequently live in a relatively computer-free world. With computers rare, there would be no need to link them together, and no need to develop the World Wide Web. Commerce, journalism, entertainment, and politics would exist under the same constraints they did in the 1930s. If we'd had a revolution in energy storage (like the pulps predicted) rather than information storage, we could zip to work with jet packs, but once we got there we'd find no cell phones, no DVD or personal video recorders, no laser printers, and no personal computers.

The field of solid-state physics, which enabled the development of these and other practical devices, is in turn made possible through quantum mechanics. While science fiction writers were imagining what the future would look like, scientists at industrial laboratories and research universities were busy using the new understanding of the quantum world to create the transistor and the laser. These basic devices form the foundation of our modern lifestyle and have transformed not just consumer electronics, but chemistry, biology, and medicine as well. All of our lives would be profoundly different if not for the efforts in the first quarter of the twentieth century of a handful of physicists trying to understand how atoms interact with light. These pioneers of quantum mechanics recognized that they were changing the face of physics, but they almost certainly did not anticipate that they would also change the future.

* * *

In this book I will explain the key concepts underlying quantum mechanics and show how these ideas account for the properties of metals, insulators, and semiconductors, the study of which forms the field of solid-state physics. I'll describe how the magnetic prop-

erties of atomic nuclei and atoms, an intrinsically quantum mechanical phenomena, allow us to see inside the human body using
magnetic resonance imaging and store vast libraries of information
on computer hard drives. The wonders enabled by quantum mechanics are almost too many to name: devices such as lasers, light-
emitting diodes, and key-chain memory sticks; strange phenomena
including superconductivity and Bose-Einstein condensation; and
even brighter brights and whiter whites!* And we'll see how the
same quantum phenomena that changed the very nature of technology in the last fifty years will similarly influence the growing
field of nanotechnology in the next fifty years.

For a field of physics that has spawned applications that have
had such a wide-ranging impact on our lives, it is unfortunate that
quantum mechanics has such a reputation for "weirdness" and
incomprehensibility. OK, maybe it is weird, but it's certainly not
impossible to understand. While the mathematics required to
perform calculations in quantum physics is fairly sophisticated,
its central principles can be described and understood without re-
sorting to differential equations or matrix algebra.

The cover of the book promised a "math-free" discussion, but
I must confess that there will be a little bit of math involved in
this presentation of quantum physics. (I hope you are reading this
at home and not standing up in the aisle at the bookstore, trying to
decide whether or not to purchase this book.) Compared to the
rigorous mathematics that underlies the foundations of quantum
mechanics, the simple equations employed here practically qualify
as "math-free." I will make use of algebraic equations no more
complex than those relating distance traveled to speed and time.
That is, if I told you that I drove at a speed of 50 miles per hour for
2 hours, you would know that I had traveled 100 miles. By arriving
at that conclusion, you have intuitively used the simple equation
distance = speed × time. None of the math that I will use here will
be more complicated than this.

While it may not be incomprehensible, quantum mechanics
does have a well-deserved reputation for being confusing. I do not
mean that the mathematics employed in a quantum description of

---

* Seriously! See Chapter 21.

nature is obscure or complex—all math is hard if you do not know how to use it, just as every language is opaque if you cannot speak it. Rather, I mean that fundamental questions, such as what happens to a quantum system when a measurement of its properties is performed, are still being argued over by physicists, nearly eighty years after first being posed. One of the most amazing aspects of quantum mechanics is that one can use it correctly and productively—even if one is confused by it.

In this book I invoke a "working man's" view of quantum mechanics that has the advantage of requiring only three suspensions of disbelief, not unlike the "miracle exception from the laws of nature" that science fiction stories or superhero comic books often implicitly employ. Some of my professorial colleagues should note—in the interest of clarity I will sidestep some of the finer points of the theory. This book is intended for non-experts interested in learning how quantum mechanics underlies many of the devices that characterize our modern lifestyle. Meditations on the interpretations of quantum theory and the "measurement problem" are fascinating, to be sure, but philosophical discussions alone do not invent the transistor.

Even keeping it simple, questions regarding the fundamental nature of matter are inescapable when considering quantum mechanics. I discuss fantastical situations such as when two electrons or atoms are so close to each other that they become "entangled" and it is actually impossible to tell them apart. I encourage you to put fear out of your mind and not shirk any necessary heavy lifting, and I'll try to hold up my end by using easily understood analogies and examples.

There are many excellent books that describe the historical development of quantum mechanics, some of which are listed in the "Recommended Reading" section. As I am not a historian of science, I will not retrace the steps of the pioneering physicists that led the quantum revolution, but will rather focus on explicating the physical principles they discovered and their applications in solid-state physics.

THE
AMAZING STORY
of QUANTUM MECHANICS

SECTION 1

TALES TO ASTONISH

Figure 1: *Cover of the August 1928 issue of the science fiction pulp magazine* Amazing Stories, *which featured the debut of "Buck" Rogers.*

CHAPTER ONE

Quantum Mechanics in Three Easy Steps

The future began twice: in December 1900, and in August 1928. On the first date, at the German Physical Society, Max Planck presented a resolution to something that would come to be called the ultraviolet catastrophe. Planck suggested that atoms can lose energy only in discrete jumps, and this new idea would tip over the first domino in a chain that by the mid-1920s would lead to the development of a new field of physics termed "quantum mechanics." On the later date, at the end of the summer of 1928, Buck Rogers first appeared in the science fiction pulp *Amazing Stories.*

With its premier issue published in 1926, *Amazing Stories* was the first magazine devoted exclusively to science fiction stories, or what publisher Hugo Gernsback called "scientifiction." The magazine's motto was "Extravagant Fiction Today . . . Cold Fact Tomorrow." Planck's breakthrough marked the dawn of a new field of science and is the province of nerds, while the appearance of Buck Rogers began the future as reckoned by geeks. (I should note that as a physics professor who is also an avid fan of science fiction and comic books, I am simultaneously a nerd and a geek.)*

Given the amazing pace of scientific progress at the end of the nineteenth century—the invention of the telegraph, telephone, and automobile had radically altered notions of distance and time, such

---

* Sorry, ladies, but I'm already married!

that, not for the last time, technology had made the world a some-what smaller place—it is perhaps not surprising that readers of *Amazing Stories* in 1928 would expect the eventual development of personal flying harnesses and disintegrator rays.

Buck Rogers's first adventure was described in Philip Francis Nowlan's novella *Armageddon 2419 A.D.*, published in that fa-mous issue of *Amazing Stories*. Anthony Rogers—he would not gain the nickname "Buck" until his appearance in a syndicated newspaper comic strip one year later—was a citizen of both the twentieth and twenty-fifth centuries. Exposure to a gas leak in an abandoned mine near Scranton induced a former army air corps officer to lapse into a form of suspended animation. Upon awaken-ing in the future, he rapidly adjusted to the new age. Nowlan's hero, catapulted into the future, was just as resourceful as Twain's Yan-kee thrust back into King Arthur's court.

Rogers, armed with the weaponry of tomorrow and a military acumen acquired during his service in World War I, joins a team of rebels fighting against the evil "Hans" invaders from Asia who had conquered America in the early twenty-second century. In fact, many of the stories published in the science fiction pulps of the 1930s and 1940s are distinguished by optimism that in the future there would be continued scientific progress coupled with pessi-mism that there would be absolutely no improvement whatsoever in international (or interplanetary) relations.

This confidence in scientific advancement, history shows, was justified, as was the expectation of continued global strife. In the pause in hostilities among European nations between the Great War and the next Great War, a revolution in physics occurred that would lay the foundation for technological innovations that would seem outlandish in the pages of *Startling Stories*. The first half of the Roaring Twenties would see the development of what would eventually be known as quantum mechanics, where the tentative guesses and first steps of Planck, Niels Bohr, Albert Einstein, and others would inspire Erwin Schrödinger and Werner Heisenberg to separately and independently create a formal, rigorous theory of the properties of atoms and their interactions with light. Their scientific papers appeared in print the same year that Hugo Gerns-back began publishing *Amazing Stories*. While quantum mechanics is not, to be sure, the last word in our understanding of nature, it did turn out to be the key missing ingredient that would enable physicists to develop the field of solid-state physics. When combined with the electromagnetic theory of the nineteenth century, quantum mechanics provides the blueprint for our current wireless world of information and communication. Scientists today, working on twenty-first-century nanotechnology, are still dining off the efforts of the quantum physicists of the 1920s.

It is plausible that the lull in global antagonisms in the brief time between the two world wars helped facilitate these advances in physics. The collaborations and interactions among scientists from Germany, France, Italy, Britain, Denmark, the Netherlands, and the United States heralded an unprecedented fertile period, which came to a close with the resumption of hostilities in Europe in 1938. Physics turned out to be in a race against history, and the pace quickened with the discovery of the structure of the atomic nucleus in the 1930s. The realization by German and Austrian physicists that it is possible to split certain large unstable nuclei, and thereby release vast amounts of energy—such that a little over two pounds of uranium would yield the same destructive force as does seventeen thousand tons of TNT—came a year before the German army marched into Poland. The quantum alliance of scientific cooperation would fracture with the formation of a geopolitical axis, and the center of gravity of physics would shift from Europe to America in the 1940s. The development of solid-state physics would have to await the end of World War II and would be carried out primarily in the United States and Britain. Unfortunately the pulp fiction writers were accurate prognosticators when they described militaristic struggles in the far future or on distant planets, suggesting that human nature evolves at a much slower pace than does technology.

Just as the hotbed of activity in physics would shift from Europe to America following World War II, the epicenter of science fiction would undergo a similar transition. Hugo Gernsback wrote in "The Rise of Scientification" in the spring 1928 issue of *Amazing Stories*, "It is a great source of satisfaction to us, and we point to it with pride, that 90 percent of the really good scientifiction authors are Americans, the rest being scattered over the world." In Gernsback's perhaps biased opinion, homegrown talent had eclipsed the seminal contributions to the genre by Jules Verne, H. G. Wells, and other European pioneers of "scientifiction."

Verne in particular is considered by many to be the "father of science fiction." He is lauded for his accurate descriptions of future technology (heavier-than-air transport, long-range submarine travel, lunar travel via rockets) as well as for his impossibly exotic locales (hollow centers of the Earth and mysterious islands). Verne's success at prediction stems from his following the same principles that guide scientific research. Whether uncovering new scientific principles or creating a new genre of speculative fiction, one must head out for uncharted terrain. One will not discover a new continent, after all, if one travels only on paved highways. As Edward O. Wilson once cautioned, for us mere mortals, who are not able to make the dramatic leaps of a Newton or Einstein, care must be taken to not metaphorically sail too far from home, in case the world really is flat. The preferred tack is to make small excursions from the known world, trying always to keep the shore in sight. Verne would frequently make reasonable extrapolations on current scientific developments and imagine a mature technology that could exist, if a few details (and perhaps a miracle exception from the laws of nature) were finessed.

A Jules Verne adventure inevitably takes place in the time period that the novel is published, and a then physically improbable mode of transportation will bring our heroes to an exotic locale. This was the format of Verne's first successful novel, *Five Weeks in a Balloon*, in which a trio of adventurers in 1863 travel to uncharted Africa, as well as his later novels *Journey to the Center of the Earth*, *20,000 Leagues Under the Sea*, *From the Earth to the Moon*, *The Mysterious Island*, and *Robur the Conqueror*. Yet in the second novel he wrote, though it was the last to be published, Jules Verne considered the most extraordinary voyage of all—to *Paris in the Twentieth Century*.

This novel marks a radical departure for Verne. Written in 1863, it describes the everyday life and mundane experiences of a young college graduate in Paris in 1960. In contrast to the optimistic view of technological wonders one associates with Verne, the novel despairs for a future world where commerce and mechanical engineering are the highest values of society, and cultural pursuits such as literature and music are disdained. So uncommercial did Verne's publisher find this manuscript decrying the triumph of commerce that he convinced Verne to lock it away in a safe. There it sat, neglected and forgotten, until the 1990s, when the safe, which was believed to be empty and whose key had long been lost, was cut open with a blowtorch, and the tome was discovered.

This short fiction certainly could never be mistaken for a typical Verne adventure tale—the protagonist is a young poet who loses his job at his uncle's bank, fails to find gainful employment, loses contact with his only friends and his young love, and ends the novel wandering aimlessly through the streets of Paris during a bitter winter storm until he passes out in the snow in a cemetery containing many famous French authors of the nineteenth century. And yet there are enough accurate descriptions of life in the next century to clearly place it among Verne's body of work. The 1863 novel describes automobiles that drive quietly and efficiently using a form of the internal combustion engine (thirteen years before Nikolaus Otto invented the four-stroke engine and more than forty years prior to the mass production of automobiles by Henry Ford), and it is suggested that the energy source involves the burning of hydrogen. Elevated trains are propelled by compressed air (while the London Underground opened the year this novel was written, elevated tracks would not see real construction for another five years); the city is illuminated at night by electric lights (Cleveland, Ohio, rather than Paris, would earn the title of first city of electric lights five years later); and skyscraper apartments are accessible by automatic elevators, again five years before the construction of the elevator in the eight-story Equitable Life Assurance Building in New York City.

Verne posited that by 1960 global communication would be an established fact and a worldwide web of telegraph wires would bring "Paris, London, Frankfurt, Amsterdam, Turin, Berlin, Vienna, Saint Petersburg, Constantinople, New York, Valparaiso, Calcutta, Sydney, Peking, and Nuku Hiva* " together. Furthermore, he described "photographic telegraphy," to be invented at the end of the nine-

---

* The largest of the chain of Marquesas Islands in what was known as French Polynesia.

teenth century, which "permitted transmission of the facsimile of any form of writing or illustration, whether manuscript or print, and letters of credit or contracts could be signed at a distance of five thousand leagues." This last had to await developments in physics more profound than pneumatic trains—for the modern fax machine is a demonstration of quantum mechanics in action!

Verne also suggested in this novel that mechanical progress would result in a military arms race that would yield such destructive cannons and equally formidable armor shielding that the nations of the world would just throw up their hands and abandon war entirely. Friends of the main character in the novel, bemoaning the loss of the honorable occupation of professional soldier, note "that France, England, Russia and Italy have dismissed their soldiers; during the last century the engines of warfare were perfected to such a degree that the whole thing had become ridiculous." Verne did accurately predict the "mutually assured destruction" theory of war ushered in by intercontinental ballistic missiles, but he underestimated the capacity of humans to find ways to wage wars nevertheless.

*　*　*

There is a deep similarity between the young physicists who developed quantum theory and the fans of the science fiction pulps of the 1920s and 1930s. Namely, they were both able to make a leap—not of faith but of reason—to accept the impossible as real and to will their disbelief into suspension.

Science fiction fans can entertain the possibility of faster-than-light space travel, of alien races on other planets, of handheld ray guns capable of shooting beams of pure destruction, and of flying cars and humanoid robots. The physicists at the birth of quantum mechanics, trying to make sense of senseless experimental data, had to embrace even more fantastic ideas, such as the fact that light, which since the second half of the nineteenth century had been conclusively demonstrated both theoretically and experimentally to be a wave, could behave like a particle, while all solid matter has a wavelike aspect to its motion.

It is perhaps small wonder that, faced with such bizarre proposals concerning the inner workings of a universe that had heretofore exhibited clockwork predictability, these scientists sought relaxation not in fantastic science fiction adventures but in the conventionality of dime-store detective novels and American cowboy motion pictures. In fact, the predictability of these western films led Niels Bohr, one of the founders of quantum theory, and his colleagues to construct theories regarding plot development in Westerns, when not grappling with the mysteries of atomic physics. In one participant's recollection, Bohr proposed a theoretical model for why the hero would always win his six-shooter duel with the villain, despite the fact that the villain always drew first. Having to decide the moment to draw his pistol actually impeded the villain, according to Bohr's theory, while the hero could rely on reflex and simply grab his weapon as soon as he saw the villain move. When some of his students doubted this explanation, they resolved the question as good scientists, via empirical testing using toy pistols on the hallways of the Copenhagen Institute (the experimental data confirmed Bohr's hypothesis).

In most discussions of quantum mechanics, at both the popular and technical levels, one typically begins with a recitation of the experimental findings that challenged accepted theories and then proceeds to describe how these data motivated physicists to propose new concepts to account for these observations. Let's not do that. In the spirit of the 1970s television detective show *Columbo*, * I'll begin with the solution to the mystery of the atom and only then describe its experimental justification.

There are three impossible things that we must accept in order to understand quantum mechanics:

Light is an electromagnetic wave that is actually comprised of discrete packets of energy.

Matter is comprised of discrete particles that exhibit a wavelike nature.

Everything—light and matter—has an "intrinsic angular momentum," or "spin," that can have only discrete values.

It is reasonable at this stage to ask: Why wasn't this brought to our attention sooner? How is it possible to live a careful and well-

---

* That is, our discussion will employ a Quentin Tarantino–esque description of quantum physics—namely, answers first, then questions.

observed life and yet never notice the particle nature of light, the wave nature of matter, and the constant spinning of both? It turns out that these are all easy to miss in our day-to-day dealings. While the human eye is physically capable of detecting a single light particle, rarely do we come across them in ones or twos. On a sunny day, the light striking one square centimeter (roughly equivalent to the area of your thumbnail) is comprised of more than a million trillion of these packets of energy every second, so their graininess is not readily apparent.

The second principle discusses the wavelike nature of matter. I show in Chapter 3 that a thrown baseball has a wavelength less than a trillionth the size of an atomic nucleus; it is consequently undetectable. The wavelength of an electron within an atom, in contrast, is about as large as the atom itself, and thus this wavelike property cannot be ignored as we seek to understand how the atomic electrons behave.

Atoms interact with light in minute quantities, and the wavelike nature of the motion of electrons in the atom turns out to be crucial to determining how it can absorb or lose the energy contained in light. Thus any model of the atom and of light that relies solely on our day-to-day experiences fails to accurately account for observation. The influence of the third principle, concerning the "intrinsic angular momentum," also referred to as "spin," is fairly subtle and comes into play when two different electrons or two atoms are so close to each other that their matter-waves overlap. This effect turns out to be rather important and is the key to understanding solid-state physics, chemistry, and magnetic resonance imaging.

While it is certainly true that these three basic principles of quantum mechanics seem weird, it is important to note that making counterintuitive proposals about nature is *not* a unique aspect of quantum mechanics. In fact, putting forth a seemingly weird idea to describe some aspect of the physical world, developing the logical consequences of this weird idea, experimentally testing these consequences, and then accepting the reality of the weird idea if it conforms to observations is pretty much what we call "physics."

Weird ideas have been the hallmark of physics for at least the past four hundred years. Sir Isaac Newton argued in the mid-1600s, in his first law of motion, that an object in motion remains in motion unless acted upon by an external force. In my personal experience, when I am driving in a straight line along a highway at a constant speed of 55 miles per hour, I must continue to provide a force in order to maintain this velocity. If I take my foot off the accelerator, I do not remain in uniform straight-line motion (even if my tires are properly aligned) but rather slow down and eventually come to rest. This is, of course, due to the influence of other external forces acting on my automobile, such as air drag and friction between the road and my tires. We do not find the effects of friction strange or mysterious, as we have had a few centuries to accept the concept of dissipative forces. These forces appear "invisible" to us, and it required tremendous insight and abstraction on Newton's part to imagine what an object's motion would be like in their absence. This strange idea of drag and frictional forces, no less counterintuitive than anything quantum theorists have suggested, applies to large objects such as people and apples.

The quantum realm is more mysterious, as most of us, aside from superheroes such as the Atom or the Incredible Shrinking Man, do not regularly visit the interior of an atom. Nevertheless, it took roughly sixteen hundred years for Newton's first law of motion to overturn Aristotle's proposal that objects slowed down and came to rest not due to friction, but owing to the fact that they longed to return to their "natural state" on the ground.

In the century preceding the development of quantum theory, physicists such as Michael Faraday and James Clerk Maxwell suggested that the forces felt by electric charges and magnets were due to invisible electric and magnetic fields. Faraday was the first to suggest that electric charges and magnetic materials create "zones of force" (referred to as "fields") that could be observed only indirectly, through their influence on other electrical charges or magnets. Scientists at the time scoffed at such a bizarre idea. To them, even worse than Faraday's theory was his pedigree: He was a self-taught experimentalist who had not attended a proper university such as Oxford or Cambridge. But Maxwell took Faraday's suggestion seriously and was able to theoretically demonstrate that visible light consists of an electromagnetic wave of oscillating electric and magnetic fields.

Changing the frequency of oscillation of the varying electric and magnetic fields yields electromagnetic waves that can range from radio waves, with wavelengths of up to several feet, to X-rays, with a wavelength of less than the diameter of an atom. Each of these forms of light are outside our normal limits of detection but can be detected with appropriate devices. The weird ideas of Faraday and Maxwell are the basis of our understanding of all electromagnetic waves, without which we would lack radio, television, cell phone communication, and Wi-Fi.

If the nature of progress in physics involves the introduction and gradual acceptance of weird ideas, then why does quantum mechanics have a particular reputation for bizarreness? It can be argued that, in part, the weirdness of the ideas underlying quantum mechanics is a consequence of their unfamiliarity. It is no less counterintuitive, in my opinion, to state that electric charges generate fields in space, and that we are always moving through a sea of invisible electromagnetic waves, even in a darkened room, than to say that light is composed of discrete packets of energy termed "photons." Phrases such as "magnetic fields" and "radio waves" are part of the common vernacular, while "wave functions" and "de Broglie waves" are not—at least not yet. By the time we are done here, such terms will also become part of your everyday conversation.*

---

* In which case, at least, you won't have to wonder why you aren't invited to more parties!

CHAPTER TWO

Photons at the Beach

Light is an electromagnetic wave that is actually comprised of discrete packets of energy.

The cover of the August 1928 issue of *Amazing Stories*, shown in Figure 1, which contained Buck Rogers's debut, featured a young man flying via a levitation device strapped to his back. While rocket packs would be labeled "Buck Rogers stuff" (and levitating belts would soon be featured in Buck Rogers's newspaper strip adventures), the cover of this issue of *Amazing Stories* actually illustrated E. E. Smith's story "The Skylark of Space." The cover depicts Dick Seaton, a scientist who is testing out a flying device that employs a newly discovered chemical. When an electrical current is passed through this substance, Element X, while it is in contact with copper, the "intra-atomic energy" of the copper is released, providing an energy source for a personal levitation belt, a spaceship (the Skylark of the title), or a handheld weapon firing "X-plosive bullets."

"The Skylark of Space" leaves vague the exact nature of the "intra-atomic energy" released by the copper when catalyzed by Element X and an electrical current. A rival scientist of Seaton's puts it as follows: "Chemists have known for years that all matter contains enormous stores of intra-atomic energy, but have always considered it 'bound'—that is, incapable of liberation. Seaton has liberated it." As chemists certainly knew, even in 1928, how to release the energy stored in chemical bonds between atoms in molecules such as nitroglycerin or TNT, the vast amounts of intra-atomic energy liberated by Element X may refer to the conversion of mass into energy through Einstein's relationship $E = mc^2$. This seems likely; when a spaceship propelled by Element X is accidently set to full thrust, the resulting acceleration becomes so great that no one on board can move to the control board to decrease their speed and the ship stops its motion only when the copper supplies are exhausted. While illustrating Einstein's principle of the interrelation between energy and mass, this scene contradicts the Special Theory of Relativity when it reveals that this uncontrolled acceleration has resulted in the ship traveling many times the speed of light. When Seaton wonders how this can be reconciled with Einstein's famous work, his companion replies, "That is a theory, this measurement of distance is a fact, as you know from our tests." Like any good scientist, Seaton agrees that observation is the final arbiter of correctness and concludes of Einstein, "That's right. Another good theory gone to pot."

The scientists in "The Skylark of Space" should not be so quick to abandon Einstein, for their X-plosive bullets of intra-atomic energy provide confirmation of another of his theories. This application of Element X, as well as the ray guns wielded by Buck Rogers, Flash Gordon, and other heroes of the science fiction pulps and comic strips, is not too far from the mark, as reflected in the first quantum principle, at the top of this chapter. As proposed by Albert Einstein the same year he developed his Special Theory of Relativity, all light consists of "bullets," that is, discrete packets of energy, termed "photons."

Now that we have the answers to quantum mechanics—what were the questions that called for these new physical principles? The ultraviolet catastrophe alluded to earlier concerned the brightness of the light emitted by an object as a function of temperature. Certain objects, such as graphite and coal dust, are black, as they absorb nearly all light that shines on them. In equilibrium, the light energy absorbed is balanced by light given off. The spectrum of light of such blackbodies, that is, how much light is emitted at a given frequency, depends only on how hot it is and is the same for metals, insulators, gases, liquids, or people if they are at the same temperature.

The theory of electromagnetic waves, developed by James Clerk Maxwell in the second half of the nineteenth century, was

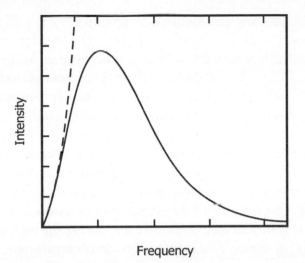

Frequency

Figure 2: *A plot of the light intensity given off from a "blackbody" object as a function of the frequency of light. The measured curve (solid line) shows that the total amount of light emitted is finite, while the pre-quantum mechanics calculated curve (dashed line) continues to rise as the frequency of light increases. That is, before quantum mechanics, physics predicted that even objects at room temperature would give off an infinite amount of light energy in the ultraviolet portion of the spectrum—a clearly ridiculous result.*

able to account for the energy emitted by a glowing object at low frequencies, such as infrared light, but at higher frequencies (above visible light) this theory predicted results that were nonsensical. Calculations indicated that the light from any heated object would become infinitely intense at high frequencies, above the ultraviolet portion of the spectrum. Thus, anyone looking at the glowing embers in a fireplace, or the interior of an oven, should be instantly incinerated with a lethal dose of X-rays. If this were true, most people would notice. This so-called ultraviolet catastrophe (which, as indicated, was a catastrophe more for theoreticians making the predictions than for anyone else) disappeared following Planck's suggestion that when the atoms in a glowing object emit light, the atoms lose energy as if they were moving down the steps of a ladder, and that those atoms must always move from rung to rung of the ladder and cannot make any other transitions between rungs. Why this would resolve the ultraviolet catastrophe, we'll explain in Section 4. For now let's focus on this "ladder" of possible energy values.

Planck firmly believed that light was a continuous electromagnetic wave, like ripples on the surface of a lake,* as theoretical considerations and extensive experimental evidence indicated. His proposal of discreteness in atomic energy loss was fairly modest (or as modest as a revolution in scientific thought can be). It turns out that while Planck justly receives credit for letting the quantum genie out of the bottle, there were other experimental conundrums waiting in the wings regarding how atoms interacted with light that would require far bolder steps than Planck was willing to take. At the same time that scientists were measuring the light given off by hot objects, giving rise to the ultraviolet catastrophe, Philipp Lenard was studying the electrons emitted by metals exposed to ultraviolet light. This led to a different catastrophe, both personal and scientific.

In the late 1800s physicists had discovered that certain materials, such as radium and thorium, gave off energy in the form of what would eventually be termed "radiation." Scientific researchers entered a "library phase," cataloging all of the different types of radiation that different substances emitted. Using the Greek alphabet as labels (α, β, γ, etc., instead of a, b, c, and so on), they started with "alpha rays," which turned out to be helium nuclei (two protons and two neutrons) ejected from atoms found near the end of the periodic table of the elements,[†] then moved on to "beta rays," (high-speed electrons), followed by "gamma rays" (very high-energy electromagnetic radiation).[‡] When William Roentgen discovered a form of radiation that would fog a photographic plate, passing through paper or flesh but not metal or bones, he termed this unknown ra-

---

* As light from the sun reaches us through the vacuum of empty space, electromagnetic waves are unique in not requiring a medium in which to propagate.

† When a nucleus emits an alpha particle, it transmutes into another element, as discussed in detail in Section 3.

‡ George Gamow, brilliant physicist and famed practical joker, once added Hans Bethe (pronounced "beta") as a coauthor of a paper he wrote with his graduate student Ralph Alpher, so that the scientific citation list of authors would read, Alpher, Bethe, Gamow.

diation "X-rays." Roentgen's discovery came before the Greek nomenclature tradition used for naming rays; he used the letter X, as it is the letter traditionally employed in math problems for the unknown quantity. (Roentgen's "X-rays" were the forerunner of many science fiction X-based characters, such as the X-Men, Professor X, Planet X, Dimension X, Element X, and X the Unknown). It was later shown that X-rays are simply electromagnetic waves—that is, light—with more energy than visible and ultraviolet light but less energy than gamma rays.

These varieties of radiation provided scientists at the end of the nineteenth century with new tools to study matter. By exposing different materials to these forms of radiation and observing their effect, they could probe the inner working of atoms. They were able, for the first time, to metaphorically take off the back plate and examine the mechanisms inside the atomic watches. Admittedly this tool was more like a hammer than a jeweler's screwdriver, but you use what you have.

Lenard was working at the University of Heidelberg and investigating the influence of light exposure on various metals. He discovered through a series of careful experiments that certain metals, when illuminated with ultraviolet light, give off beta rays, that is, electrons. It would turn out that the electrons he was observing originated from the sea of electrons that explain why all metals are good conductors of heat and electricity. This, in hindsight, is not that surprising. Light carries energy, and when an object absorbs energy it warms up. Some of the excess energy in the metal can be transferred to the electrons, and if they are sufficiently energetic they can fly free, not unlike the energetic water molecules leaving the liquid surface of a hot cup of coffee and forming a cloud of steam above the mug. Philipp Lenard set about systematically investigating how the number of electrons emitted from a given metal and their speed depended on the frequency and intensity of the ultraviolet light he used. Here the troubles began.

Imagine a metal as a sandy beach at the ocean's shore, and the electrons in the metal as small pebbles randomly scattered along the beach (see Figure 3). The ocean waves crashing onto the beach can be considered the ultraviolet electromagnetic waves shining on

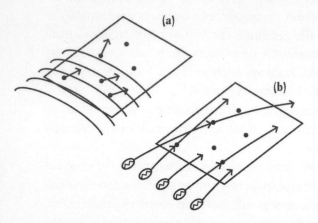

Figure 3: *Cartoon sketch of pebbles on a beach, pushed up toward the top of the beach by either ocean waves (a) or by photon bullets (b).*

the metal in Lenard's experiment. This allegorical beach has a gentle slope away from the water's edge, so that one must do work against gravity to push the pebbles away from the shoreline. When the pebbles have advanced up the beach all the way to the boardwalk, we'll consider them free, and as our stand-ins for electrons in a metal, they will represent those electrons that are ejected from the solid by the ultraviolet light in Lenard's experiment. In this analogy it is reasonable that the energy responsible for picking up the pebbles and moving them up the beach comes from the water waves. The bigger the wave, the more energy the pebbles will have. The more frequently the waves arrive, the greater the number of pebbles will be advanced. If the waves have small amplitudes, one might have to patiently wait for several wave fronts to transfer sufficient energy to the pebbles before they can move up the beachfront. Perfectly reasonable—except this was not what Lenard observed.

He found that the energy of the electron leaving the metal did not depend on the intensity of the light. Raising the light intensity did not affect the electron's speed, though it did increase the number of electrons emitted per second. But the number of electrons ejected per second was supposed to depend on the frequency of the waves, that is, how many wave crests arrived per second, and not their amplitude. Moreover, there was a threshold for electron emission—if the frequency of the light were below a certain value (which varied for different metals), then no electrons came off, regardless of how bright the light. In the water analogy, this would suggest that if the number of wave peaks per second were below a given value, then even a tsunami would not push the pebbles up the beach. What did govern whether or not the electrons were emitted was the frequency of the light. For the beach analogy, this would be as if once the number of wave peaks per second rose above a certain threshold, then even a very low-amplitude, gentle wave would promote pebbles up to the boardwalk, just as long as the time between crests was short enough. Finally, if the frequency of the light was above this threshold, then electrons came off instantly, with no time delay, no matter how low the light's intensity. These experimental results were a challenge to understand within the context of light being a continuous electromagnetic wave, but we already know the answer to this question, stated at the start of this chapter. Light is not a continuous wave but is composed of individual energy bullets, termed "photons."

How does light actually consisting of discrete packets of energy explain Lenard's experimental results? What we took to be a continuous, uniform series of waves washing up onshore, gently pushing the pebbles up the beach, is actually comprised of bullets fired at the pebbles by a machine gun sweeping back and forth across the beach. With light being a collection of photons, the brightness of the light is determined by the number of photons passing through a given area per second. For our machine gun, this is equivalent to the rate at which bullets are fired; that is, more bullets per second leads to more intense illumination. One bullet a second is a weak light source, while a million shots per second is much brighter. The energy of the photons would be reflected in our analogy in the speed of the bullets (let's suppose for the sake of argument that we have a magical machine gun that provides independent control over the speed of the bullets it fires).

It turns out that if you imagine the light striking the metal as analogous to a spray of machine-gun bullets rather than continuous ocean waves, Lenard's results are completely reasonable. He found that the brighter the light, the more electrons were emitted from the metal. In the machine-gun analogy, brighter light means more bullets per second, which will push more pebbles per second.

There will be no time lag between light coming into the metal and electrons coming out, for once a bullet hits a pebble, and it has sufficient energy to knock it up the beach, then of course the effect will be instantaneous. The faster the bullets are traveling, the more energy will be imparted to the pebbles, and the faster they will move up the beach. This corresponds to saying that the greater the energy of each light photon, the more kinetic energy the ejected electrons will have. If the speed of the bullets is too low, they may move the pebbles a little bit but not knock them significantly up the beach. So the threshold effect Lenard observed is also explained. The only snag is that Lenard did not control the energy of his light to vary the energy of the emitted electrons, but rather its frequency. And here we come to the personal catastrophe for Lenard, for the resolution of this last remaining puzzle would cause him distress of a decidedly nonscientific nature. (More on this in a bit.)

If light does indeed consist of discrete packets of energy, what determines the energy of each packet? Planck's solution to the spectrum of glowing, hot objects was to propose that atoms could lose energy only in finite jumps. In order to get the equations to work out, he assumed that the energy of the jump was proportional to the frequency of the light. That is, the bigger the frequency, the larger the energy of the "quantum step." Again, I'll explain why this worked later on. For now the important point is that if the energy is proportional to the frequency, then we can say that the energy is equal to the frequency when multiplied by a constant.

We often deal with situations involving simple proportions, such as the relationship that the longer you drive your automobile at a constant speed, the greater the distance you travel. But you do not measure distance in hours, so to figure out how far you have driven, you need to multiply the hours in the car by a "constant of proportionality," namely, your constant speed (say 60 miles per hour). Then the product of the time driven (2 hours) and the uniform speed (60 miles per hour) will determine the distance traveled (120 miles). In the same way, the energy of the photon is proportional to the frequency, so that when the frequency is multiplied by a constant of proportionality, it is converted into a quantity of energy. Planck used the letter h to represent this constant of proportionality, and everyone who has followed has stuck with that convention, so that h is referred to as "Planck's constant." The equation for the energy lost by an atom in a glowing object by the emission of light is as follows:

$$Energy = h \times (Frequency)$$

Let's plug some numbers into this simple equation, which is mathematically no different from "distance = (speed) × (time)." One way to measure energy is in a unit termed Joules, named after James Joule, a Scottish physicist who demonstrated the clear equivalence between heat and mechanical work, thereby providing a foundation for the field of thermodynamics. For point of reference, a major-league baseball thrown at 60 miles per hour has a kinetic energy of 53 Joules, while an automobile traveling at 60 miles per hour has a kinetic energy of 600,000 Joules. Frequency is a measure of how many times a periodic function repeats a complete cycle in a given unit of time and is most naturally measured in terms of number of cycles per second. A child's playground swing that takes 2 seconds to go all the way back and forth will complete only one half of a cycle in a second, so it has a frequency of 1/2 cycle/second. A much faster playground swing that goes back and forth in 1/10 of a second will complete ten loops in 1 second, and thus has a frequency of 10 cycles/second. Visible light has a frequency of one thousand trillion cycles per second. In order to fit his equation for the spectrum of light given off by a hot, glowing object to the experimentally measured curves, Planck had to set the value of his constant h to be $h = 660$ trillionth trillionth trillionth of a Joule-sec, which may seem very small but is in fact very, very, very supersmall.

Planck argued, when justifying his proposed equation, that atoms could lose energy only in finite steps. The closest these energy levels could be was $E = h \times f$. For light with a frequency of a thousand trillion cycles per second, this equation gives a spacing between adjacent energy levels of 0.66 millionth trillionths of a Joule. If you compare this energy of a photon to the 50 Joules of kinetic energy of a fastball, you can see that the separation is so small compared to the energy scale of our everyday activities that it's nearly impossible to observe. Consequently, to take a quantum jump involves a very, very tiny baby step and not some enormous leap. But the size of the step is, in fact, irrelevant, as the important and revolutionary concept introduced by Planck is that there *is* a step.

As mentioned, Planck was actually somewhat conservative in his quantum hypothesis. For him light was still a continuous electromagnetic wave, like the metaphoric ocean waves washing onto the shore. While he argued that atoms could lose energy only in discrete steps, he was not so bold as to suggest that when they did so the light emitted was also a discrete packet of energy. But as the humorist James Thurber once wrote, "Fools rush in where angels fear to tread. And all the angels are in heaven, but few of the fools are dead." Perhaps Planck's hesitancy to extend the quantum graininess to the light itself came from the caution of age. He was a grand old man of forty-two, after all, when he developed his quantum hypothesis. It was left to a younger man of twenty-six to suggest that Lenard's results could be qualitatively and quantitatively accounted for if light itself were discrete packets, the machine-gun bullets in our beach metaphor, where the energy of each bullet is determined by $E = h \times f$. That young man was Albert Einstein.

When Einstein wrote his paper on the "photoelectric" effect, for that is the name used to describe Lenard's experiment, he was an underemployed twenty-six-year-old patent clerk, third class. That would soon change, for Einstein's paper was published in 1905, the same year he published his Special Theory of Relativity, followed by a paper describing the equivalence between energy and mass—$E = mc^2$—and two other papers on an atomistic explanation for Brownian motion and diffusion processes that would have cemented his reputation as a theoretical physicist of the first order even if he'd done nothing else. Within a few years of the papers' publication, Einstein would be offered professorships and honors. Most scientists would be thrilled to have their work permanently associated with Einstein, but not Lenard, for one simple reason. Einstein was Jewish, and Lenard was a rabid anti-Semite, to such an extent that Adolf Hitler named Lenard chief of Aryan physics.

Thus did his experiments on the influence of ultraviolet light on metals cause a personal catastrophe for Lenard—he spent such effort denouncing Einstein and his interpretation of the photoelectric experiments that his own scientific reputation was ruined, and he is now remembered as much for his bigotry as for his talent as an experimentalist.

Fortunately for Einstein, Lenard was not the only physicist who strongly disagreed with the hypothesis that light was comprised of discrete packets of energy. Nearly all physicists, including the American physicist Robert Millikan, were convinced that light was a continuous wave, and that Einstein's suggestion (the title of his original paper described his proposal as a "heuristic viewpoint"— which is fancy talk for "not a rigorous solution, but I should get the credit if it turns out to be right") could not possibly be correct. I say "fortunately for Einstein," for he did indeed turn out to be right, and it was Millikan who proved it. Millikan was one of the most careful and gifted experimentalists of his day, and he spent ten years trying to show that Einstein was wrong. What he wound up showing, in fact, was that the only possible explanation for the

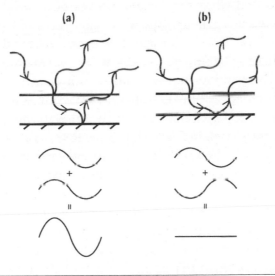

Figure 4: *Sketch of a light wave reflecting from the top and bottom surfaces of an oil slick. In (a) the wavelength and thickness of the slick result in constructive interference of the wave reflected from the top surface, and the wave that travels through the slick reflects from the bottom surface and then exits the slick. For constructive interference the light would be very bright when viewed from the top. In (b) the wavelength and thickness result in destructive interference, in which case no light would be observed from the top surface.*

photoelectric effect was Einstein's hypothesis. Even though he still believed that Einstein's photon idea was crazy, he stood by his data, which not only unambiguously supported Einstein's claim but also won Millikan the Nobel Prize. The best advocate for your position is someone who initially doubts but is converted by overwhelming evidence. Einstein's heuristic viewpoint is now everyone's viewpoint, and Einstein received the Nobel Prize in Physics, not for his work on relativity or $E = mc^2$, but for his theory of Lenard's photoelectric effect and the introduction of the concept of the light photon (though Gilbert Lewis was the first to use the term "photon," in 1926).

Now, for a very long time, a lot of smart people firmly believed that light was a wave, based on a large body of compelling experimental evidence. To cite just one manifestation of the wavelike nature of light, consider the rainbow of colors one sees on an oil slick following a rainstorm. Anyone who lives in a city, or has a messy driveway, may be familiar with the spectrum of light reflected from a thin layer of oil when the ground is wet from a good soaking. In this case the oil is repelled by the water and sits as a thin, freestanding slab that may be only a fraction of a millimeter thick. Rarely will the surface be atomically uniform, so the thickness of the oil slick will vary from location to location. Some of the light striking the oil slick reflects from the top surface, while some penetrates through the oil and then bounces off the oil-water interface. As illustrated by the sketch in Figure 4, if the thickness at one particular spot on the oil slick is exactly equal to one-fourth* of the wavelength of one particular color of light, then the light reflected from the top and the light that passed through the slick and bounced back out will be exactly in phase and will add up coherently. The light at that color will thus be particularly bright. Just as two pendulums, if they are swinging side by side with exactly the same frequency and the same phase (so that they both are at the top of the arc or both at the bottom of the swing at the same time), are said to be

---

* For an oil slick on water, the thickness needs to be one-fourth the wavelength of light for constructive interference, while for a glass slab with air above and below, the constructive interference criteria call for the thickness to be one-half the wavelength. The difference, involving phase changes at the top reflection surface, is not important for the discussion here.

coherently oscillating, the light that is coherently reflected will add up and give a much more intense color. All other colors will have wavelengths that are not exact multiples of the oil-slick thickness at that location, and they will not add up constructively. This familiar phenomenon is termed "interference" and is a signature property of waves.

We therefore must figure out why light exhibits phenomena that indicate that it is a wave, when it is in fact comprised of photons. It is tempting to argue that the wavelike properties of light are some sort of collective behavior when many photons interact with one another. Oh, if it were only this simple. Consider the following situation: If one sent not a continuous beam of light at the oil slick described above, but single photons, one at a time, then the photons would reflect from the slick, from either the top or bottom surface, and be detected by some sort of light sensor. We would see a single flash at a particular location on our detector when each photon had reflected from the slick. When many such photons had been reflected, the resulting pattern of flashes of light observed would be an interference pattern, identical to what we would see if a continuous beam had been used. That is, even though the photons saw the slick individually, they reflected in such a way that when added together, they yielded a wavelike constructive interference pattern.

Technically, a photon is defined as a quantum of excitation of the radiation field. Well, that certainly clears that up! For our purposes, we will simply accept the notion of photons as discrete entities that move at the speed of light, that have a definite energy (and hence frequency, through $E = h \times f$), definite momentum (and hence wavelength, through the relation wavelength = speed of light/ frequency—which we will discuss later on), and definite intrinsic angular momentum (the "spin" of a photon = $h/2\pi$, measured in the direction it is moving). The photon does not spread out as it travels, the way a wave on the ocean's surface does, but propagates unchanged until it interacts with matter or other photons. There are no simple or satisfying answers to such questions as how big the photon is, or whether it is a wave or a particle. If you find this confusing—join the club! It's a big club with some rather distinguished members. The president of this club would be the scientist who introduced the photon concept in the early days of quantum mechanics. As Albert Einstein reflected, "All the fifty years of conscious brooding have brought me no closer to the answer to the question: What are light quanta? Of course today every rascal thinks he knows the answer, but he is deluding himself."

CHAPTER THREE

Fearful Symmetry

Matter is comprised of discrete particles that exhibit a wavelike nature.

Readers of the February 1930 issue of *Science Wonder Stories* were treated to thrilling tales of the "Streamers of Death" and "A Rescue from Jupiter"; they traveled to "The Land of the Bipos" and visited "The World of a Hundred Men." The cover features a scene from the "Bipos" yarn. Two robbers who have broken into the home laboratory of a Dr. Sanborn, who was experimenting on methods to send living beings to another world (whether in this universe or an alternate one is never made clear), have been trapped in a large glass device. This cylinder, large enough to hold two grown men, is described in the story as a "cathode ray tube"—though its appearance is quite different from the cathode-ray tubes one finds in older-model television sets. Sanborn is shown moments before throwing a switch that will convert the two thieves into electricity. They will then travel at the speed of light to the land of the Bipos, where they will be reassembled into their human form. Bipos, apparently, are a race of intelligent three-foot-tall penguins. The means of transportation appears to be an early ancestor of *Star Trek*'s famed transporter. That Sanborn was able to construct such a fantastic scientific marvel, with no outside assistance and using his own financial resources, is perhaps not so surprising once we discover that in his day job Dr. Sanborn is . . . a druggist!

Science Wonder Stories was not devoted solely to fantastic scienctifiction but also featured descriptions and discussions of real-

Figure 5: Dr. Sanborn about to test his home-made transporter device (that looks like an over-grown vacuum tube), which will send two intruders to the Land of the Bipos in 1930's Science Wonder Stories.

world current scientific advances. This particular issue contained a "Symposium," in which an essay on the question "Can Man Free Himself from Gravity?" was followed by letters from knowledgeable experts. The short essay by Th. Wolff of Berlin was translated for the pulp from the original German. Wolff tantalized readers with a report of an American physicist, Charles Brush, who claimed to have discovered a material made up of silicates (the exact composition known only to Brush) that exhibited an acceleration due to gravity of only 9.2 meters per second per second, rather than the larger value of 9.8 meters per second per second that all normal matter experiences. "If true, this would be a fine achievement," allowed Wolff, for "by increasing the valuable property of these mysterious substances one might perhaps attain approximate or even complete freedom from gravity. Let us wait for it!"

But Wolff did not think we should hold our breaths while waiting, for he went on to correctly point out that such a material would represent an "irreconcilable contradiction" to the Newtonian law of gravity, which indicates that the acceleration of a falling object is the same for all matter, regardless of composition. Brush's report, Wolff informed readers, "must with absolute assurance be relegated to the realm of fiction. If there were exceptions and deviations from the general law of gravity, these would certainly have appeared before now in manifold and various ways, and it would not need the discovery of mysterious substances to bring them to our knowledge." So much for flying cars—even back in 1930! But then Wolff goes too far—and dismisses space travel when he incorrectly calculates that the chemical fuels of the time would limit any rocket ship to heights no greater than 400 kilometers above the Earth's surface, a mere fraction of the 384,000 kilometers from the Earth to the moon.

This last point was challenged in letters from members of the *Science Wonder Stories* Board of Associate Editors, notably Robert H. Goddard of Clark University in Worcester, Massachusetts. Goddard pointed out that in 1919 he had authored a scientific publication in the Miscellaneous Collections of the Institute (namely, the Smithsonian Institute, which was funding his rocket research), stating that a multistage rocket, essentially of the design employed by NASA fifty years later, would indeed be able to exceed this 400-kilometer limit. Thus, while hopes of flying cars and perpetual motion* were dashed, the promise of rocket trips to the moon and beyond were affirmed in the science fiction pulps.

Goddard was an early example of a prominent scientist whose research would inspire many science fiction tales and whose choice of field and research subject was, in turn, inspired by science fiction. In a fan letter sent to H. G. Wells, the sixteen-year-old Goddard extolled the influence that reading *The War of the Worlds* had on him, such that no more than a year later, he "decided that what might conservatively be called 'high altitude research' was the

---

* Goddard correctly pointed out that any gravity screen as suggested by Brush would enable one to lift a large mass with little effort. Upon removal of the gravity shield, the mass would then fall as any normal weight and thus could provide a work output greater than the energy required to lift the mass, thereby violating the law of conservation of energy.

most fascinating problem in existence." Goddard was not the first scientist, of course, to find a muse in science fiction. Hermann Oberth, the Transylvanian-born scientist who is considered the "father of modern rocketry," had an encounter at age eleven with Jules Verne's *From the Earth to the Moon* that set the trajectory of his scientific career. Both Oberth and his pupil Wernher von Braun would serve as technical advisers for *Woman in the Moon,* a 1929 Fritz Lang science fiction motion picture that featured the first countdown to launch a rocket, in film or in the real world.

Real science, as opposed to fiction, was also imparted in *Science Wonder Stories'* regular features "What Is Your Science Knowledge?," "Science Questions and Answers," and "Science News of the Month." Here, in this latter section, a brief item entitled "Electron Found to Have Dual Character" read, in its entirety:

> G. P. Thompson, British scientist, has made a new discovery in the field of physics. He states that the electron acts like a flying particle and also behaves like a wave. He rolled gold, nickel, aluminum and other metals, each to about one-tenth the thickness of gold leaf, and shot electrons through them. After passing through the films the electrons came in contact with a photographic film, and were recorded as concentric circles and other circular patterns.

If the magazine had contained a detailed description of the chemical composition of an actual antigravity shield, it would not have presented a more profound or revolutionary report than this brief blurb regarding the electron's "dual character."

* * *

The second quantum principle listed at the top of this chapter states that, just as there is a particle aspect to light, there is a corresponding wavelike nature to matter. Unlike the case of the photoelectric effect in the last chapter, this strange symmetrical hypothesis about the nature of matter was not proposed in order to resolve a mysterious experimental observation that contradicted expectations of classical physical theory—but was suggested precisely because it was a strange symmetrical hypothesis.

In 1923, Prince Louis de Broglie (yes, he actually was a French prince as well as a physicist), struck by the counterintuitive suggestion that light was comprised of corpuscular particles, proposed that there was a wave—originally termed a "pilot wave"—associated with the motion of real particles, such as electrons, protons, and atoms. De Broglie had an answer for why this "pilot wave" had not been previously observed—its wavelength varied inversely with the momentum of the moving object, so the larger the object (which is easier to observe), the smaller the wavelength of its pilot wave.

How to test the proposal that there is a wave associated with the motion of matter? As mentioned in the last chapter, interference effects, such as when white light creates a spectrum of reflected colors from an oil slick suspended on a wet surface, are an excellent test of the existence of waves. To recap, when the thickness of the slick is exactly equal to specific fractions of a given color's wavelength, the waves corresponding to this color reflected from the top and those that have traveled through the slick, bounced off the bottom, and passed again through the slick and exited from the top surface add together coherently. When this happens, the color is brighter to us due to this constructive interference. Other waves corresponding to other colors at this location add up incoherently, out of phase, and the net effect is that from the white light shining on the oil, one color is primarily reflected from the slick from a given point on the slick. As the thickness of the slick can vary from point to point, we observe different colors across its surface.

The thickness of an oil slick can be several thousand nanometers (one nanometer is approximately the length of three carbon atoms, stacked one atop the other), while the wavelength of visible light ranges from 650 nanometers for red light to 400 nanometers for violet light. Thus, only very thin oil slicks, whose thickness is no more than a few times the wavelength of light, exhibit the interference pattern described above (if the slick is too thick, then the light traveling through the oil has too great a chance to be absorbed and won't make it back through the top surface). If we want to use a similar interference effect to verify the wavelike nature of the motion of matter as proposed by de Broglie, we first need to know how large or small the "matter wavelength" will be. De Broglie proposed that the connection between the wavelength of the "pilot wave" for any moving object and its momentum is given by the following expression:

$$\text{Momentum} \times \text{Wavelength} = h$$

This equation indicates that the larger the momentum, the smaller the wavelength. The product of the two quantities is a constant, and de Broglie suggested that it should be Planck's constant. Again, this equation is mathematically no different from the relationship described in the last chapter connecting distance traveled and time driving, that is, distance = (speed) × (time). In order to determine how long a car trip to Chicago from Madison, Wisconsin, may take, we note that the distance is a constant, approximately 120 miles, and not open to alteration. If our average speed is 60 miles per hour, then this equation indicates that the trip will last 2 hours. A slower speed will lead to a longer trip, and to shorten the trip to 1 hour, we must look to a speed of 120 miles per hour.* In principle, the trip may last as short or as long as we like, as long as we vary our average velocity so that, when multiplied by the travel time, it yields a distance of 120 miles.

The momentum of an object is defined as the product of its mass and its velocity. The bigger an object, the more momentum it has at a given speed, and the harder it would be to stop. Which would you rather have collide with you: a linebacker or a ballerina, both running at the same speed? If we use the mass and speed of a major league fastball in de Broglie's equation above, we find that its de Broglie wavelength is smaller than a millionth trillionth of the diameter of an atomic nucleus. There is no structure that can be conceived of that would exhibit interference effects of a baseball.

One way to increase the size of the de Broglie wavelength is to decrease the momentum of the object, as their product is a constant, and the simplest way to do that is to consider smaller objects. That is, the smaller the object, the lower its momentum (just as the ballerina has a smaller momentum than the football player), and consequently the larger its de Broglie wavelength. An electron obviously has a much smaller mass than a baseball, and a correspond-

---

* We must also look for the police!

ingly smaller momentum. Even for an electron traveling at a speed of nearly 1 percent of the speed of light, its momentum is a trillion trillion times smaller than the baseball's, and its corresponding de Broglie wavelength is a trillion trillion times larger. For just such an electron the de Broglie wavelength turns out to be about one-fourth of a nanometer, or roughly the diameter of an atom. In order to observe interference effects that would reflect the wavelike nature of matter, we would thus need to send a beam of electrons at an "oil slick" that is only a few atoms thick. That's still pretty small, but fortunately nature provides us with just such "slicks"— we call them crystals.

Any solid the size of a sugar cube, such as a sugar cube, contains a little under a trillion trillion atoms. How these atoms are arranged, their chemical composition, and the nature of the connections to their neighbors determines whether the solid in question conducts electricity and is shiny (that is, reflects light), like a metal, or does not conduct electricity and is transparent to visible light, like a diamond. Consider a carbon atom, which chemically prefers to have four chemical bonds. There are many different ways a collection of carbon atoms can chemically bond to one another, and if one brings the atoms together in a haphazard, random manner, one such resulting configuration is "soot."* An alternative bonding scheme would have each carbon atom located carefully in relation to its neighbors, so that all four chemical bonds for each carbon atom have their ideal strength and location. One such uniform, periodic arrangement of carbon atoms is termed "diamond." Chemically diamond and soot are identical as they both consist of carbon atoms bonded to each other, and yet they have very different structural properties (diamond is hard, while soot is soft), electrical conduction (soot is a pretty good conductor of electricity, while diamond is an excellent insulator), optical characteristics (soot absorbs visible light, which is why it appears black, while diamond is transparent in the visible portion of the spectrum), and financial

---

* While technically "soot" refers to particulates formed from the incomplete combustion of fuels such as coal, oil, or wood, and is mostly carbon but may contain other elements depending on the nature of the burning material, here I am using the term as a shorthand for "amorphous carbon."

(diamond is expensive precisely in proportion to its scarcity—and don't try to give a soot ring to your beloved*). If soot and diamond are identical chemically, then all of these differences must be due to the arrangements of the carbon atoms in the two substances. A deep understanding of why carbon atoms would form certain types of chemical bonds in one circumstance and very different bonds in another would not arrive until the full, formal theory of quantum mechanics was developed by Schrödinger and Heisenberg. I describe how quantum mechanics accounts for all of chemistry later on— for now we are interested in the fact that for certain solids the atoms are arranged in periodic arrays, like the oranges stacked in a grocery store display, which enables large-scale three-dimensional uniform crystalline solids.

These crystalline arrangements of atoms can be used as atomic-scale "oil slicks" for interference experiments, as shown in Figure 6, providing uniform layers that reflect electron beams striking them, with each layer being one atom thick, which is just the right fraction of the length of the de Broglie wavelength of our electrons. Thus, if we send in a beam of electrons traveling at the right speed, their momentum will be such that their corresponding "pilot wave" will have a de Broglie wavelength commensurate to the spacing between atomic layers in our crystal. The incoming electrons will be repelled by the electrons around each atom in the crystal—as identical electrical charges experience a repulsive force. As any given collision between the electron beam and the crystal's electrons is random, one would expect that the intensity of scattered electrons would be fairly uniform, regardless of the direction one looks. But thanks to quantum mechanics, this is not what is seen.

Just as in the case of the light scattered from an oil slick, where all colors are present in the incident white light, but only certain colors constructively interfere, one finds that the intensity of scattered electrons is not uniform. Rather, there are regions where a high intensity of scattered electrons are found, and other regions devoid of electrons, with a pattern exactly as one would expect for interfering waves, rather than colliding particles. Figure 7 shows

---

* Certainly not if you want them to remain your beloved!

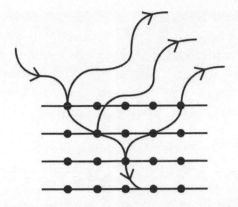

Figure 6: *Cartoon sketch of de Broglie matter waves for electrons scattering from the planes of atoms in a crystal. If the separation between atomic planes in the solid is commensurate with the de Broglie wavelength of the electrons, then interference of the scattered electrons will be observed. The intensity of electrons will be high in directions where the matter waves constructively interfere and there will be no observed electrons in directions for which destructive interference occurs.*

strikingly similar interference patterns when green light from a laser passes through a fine metal mesh and when an electron beam passes through a graphite crystal. The wavelength of green light is much longer than that of the electron beam's de Broglie waves, so the spacing between wires in the metal screen is correspondingly larger than the separation between atoms in the carbon crystal. The intensity of scattered electrons from uniform layers of atoms in a crystal (when the electrons have a suitable momentum so that their de Broglie wavelength is equal to the spacing between crystal planes) displays an identical interference pattern as is seen when X-rays, that also have a wavelength of the same size as the atomic spacing, are reflected from the same crystal. This interference pattern holds not only for reflected electrons but also for those passing through the thin crystal, as in Thompson's experiments summarized in the February 1930 issue of *Science Wonder Stories*.

As in the case of photons, described in the previous chapter, this interference effect is not a result of large numbers of electrons behaving in a collective fashion like a wave. Consider the electrons passing through the crystal in Figure 7, detected by striking a chemically coated screen that emits a flash of light whenever an

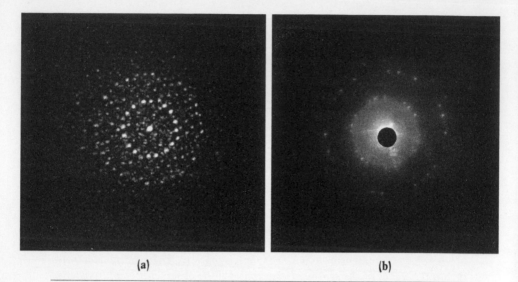

(a)   (b)

Figure 7: *Examples of light diffraction (a) and electron diffraction (b). The image on the left is obtained by passing a green laser light through a fine-mesh metal screen (not unlike a screen door) and shining the light on a wall several feet from the screen. The light scattering from the metal wires, arranged in a periodic array, leads to a symmetric constructive (bright-green spots) and destructive (dark regions) interference pattern. On the right an electron beam in a cathode ray tube passes through a graphite crystal. The momentum of the electrons is chosen so that their de Broglie wavelength is on the order of the spacing between atoms in the crystal. The atoms in the crystal scatter the electrons in a similar manner as the wire mesh does to the laser beam.*

electron strikes it. You're probably familiar with this—it's an old-style cathode-ray television tube. (Modern flat-panel liquid crystal display models work differently.) Decreasing the current of the incoming beam of electrons striking the crystal, we can arrange it so that only one electron strikes the crystal every few seconds. We would then not see a full interference pattern, but a series of individual flashes of light on the TV detector screen. The more electrons we send in, the more flashes of light. If we recorded the location of each flash and at the end of the day added them all together, instead of a uniform coverage over the screen—as would be expected if the hard-sphere electrons collided with the electrons in the crystal's atoms, sending them randomly in all directions—we see an interference pattern as in Figure 7.

X-rays (more about these later as well) correspond to electromagnetic waves with a wavelength roughly equal to an atomic diameter, just as for the electrons we considered in the scattering experiment. The X-rays scatter from the electrons in the crystal's atoms, though the mechanism is a little more involved than simple electron-electron repulsion. But we can decrease the intensity of the light, so that one X-ray photon strikes the crystal every few seconds as well. Here again, a detector screen will record distinct flashes of light, and when all the flashes are added together from the scatter of many photons, the observed interference pattern is identical to that found using electrons.* The "dual character" symmetry between particles and waves holds for both matter and light. This, we will see, is truly the most amazing science story of the twentieth century.

---

* There may be a slight difference, owing to technicalities. The electrons scatter most strongly from the top few atomic planes, while the X-rays can penetrate deeper into the crystal. This is because the electron-electron repulsion that governs the electron scattering is much stronger than the photon-electron interactions. If the crystal structure of the top surface differs from that of the bulk crystal, a different pattern may be observed. But this is a detail that does not affect the basic point of wave-particle symmetry.

CHAPTER FOUR

It's All Done with Magnets

Everything—light and matter—has an "intrinsic angular momentum," or "spin," that can have only discrete values.

After years of unsuccessful attempts to land a newspaper distribution deal for a comic strip featuring their creation Superman, Jerry Siegel and Joe Shuster eventually sold their story, along with the rights to the character, to the comic book publisher National Allied Periodical for $130—a nice sum in 1938, but of course a pittance compared to what the character would soon be worth. Debuting in *Action Comics* # 1 in June 1938, the Man of Tomorrow would soon be selling millions of comics per month and starring in live-action and animated movie shorts, an extremely popular radio show, and a syndicated newspaper comic strip.

No doubt one of the strong appeals of Siegel and Shuster's comic-book creation is the fantasy that mild-mannered Clark Kent, dismissed and underestimated by all, is in reality the most powerful person on the planet. As Jules Feiffer argued in *The Great Comic Book Heroes*, Bruce Wayne must don his Batman costume in order to become the Caped Crusader, and Lamont Cranston his cloak, slouch hat, and red scarf to fight crime as the Shadow, but Superman is who he is. When he wakes up in the morning, he is Superman, and Clark Kent is the disguise he elects to wear. In essence, Kent is a representation of how Superman views us: weak, bumbling, inept. Superman is the iconic role model for those who feel that the world does not see their true, hidden essence.

One of the surprising discoveries of quantum mechanics, described in the principle at the start of this chapter, is that electrons,

Figure 8: In the 1960s, Dick Tracy comic strips predicted a future in which we traveled via personal flying garbage cans levitated by the power of magnetism.

protons, and neutrons, the building blocks of atoms, also have a secret identity. While physicists in the 1920s knew them to be mild-mannered subatomic particles, characterized by their mass and electrical charge, it turns out that they would soon discover that the particles possessed a hidden characteristic, a superpower if you will, that is termed "spin."

It was proposed in 1925 that every fundamental particle behaves as if it is a spinning top,* rotating about an internal axis—and this holds not just for matter, but for photons as well. This rotation is not associated with the "orbital motion" of electrons around the nucleus in an atom (Schrödinger would eventually show that the picture of electrons circling around the positively charged nucleus, a neat analogy to the planets orbiting the sun in our solar system, is not technically accurate). This internal rotation is present even if the subatomic particles are in free space, not bound in an atom or molecule.

This built-in rotation is called "spin," for it is as if the electron is rotating about an axis passing through the particle itself—similar

---

* We'll soon see that, although it is a useful metaphor, we should not take the "spinning top" picture literally.

to a twirling ballerina. The fact that all subatomic particles have an internal rotation turns out to be pretty important. Without accounting for the spin of electrons we cannot make sense of chemistry and solid-state physics. One characteristic of all atomic particles, associated with their internal spin, is that electrons, protons, and neutrons all have an internal magnetic field that has nothing to do with the magnetic field generated by an electrical current. It is through this magnetic field that this "power," that is, spin, first revealed itself to the world.

Science fiction pulps often cited magnetism as the basis for a variety of technological wonders. Magnets were frequently called upon as a catch-all explanation for levitating heavier-than-air ships, while "reverse magnetism" was often invoked for force beams or other offensive weaponry. Readers of the daily newspaper's comic-strip page have known since the mid-1960s that personal flying devices would someday be a reality, thanks to magnetism. Figure 8 shows a panel from a 1960s Dick Tracy comic strip, where Tracy, in silhouette, and his partner, Detective Sam Catchem, are able to

Figure 9: *Angular momentum was a frequently invoked physics principle for futuristic weapons of war, as shown in the cover of the April 1930 Air Wonder Stories.*

scout for criminals using magnetized flying garbage cans. (Tracy is also carrying on a conversation using a "two-way wrist radio," an early form of the cell phone.) Magnetism was expected to usher in the world of tomorrow—as the panel reproduced in Figure 8 promised, "The nation that controls magnetism will control the universe."

Similarly, spinning is also a hallmark of the flying saucers and futuristic weapons (or sometimes both, as shown on the cover of the April 1930 issue of *Air Wonder Stories* in Figure 9). It would have to wait for the full relativistic form of quantum mechanics, developed by Paul Adrien Maurice Dirac in 1928, to show the fundamental connection between internal rotation and magnetism.

The third quantum principle states that everything, matter and photons, has an internal rotation about an axis that passes through the object, like a twirling figure skater. For ordinary matter, there is only one question about the rotation—clockwise or counterclockwise? In the previous chapter we discussed linear momentum defined as the product of an object's mass and velocity. Since the objects in Chapter 3 were moving in straight lines, we could employ the linguistic shortcut and just refer to it as "momentum" rather than the more accurate term "linear momentum." The greater an object's momentum, the harder it is to change its motion. A baseball thrown at 100 miles per hour has more momentum than one thrown at 1 mile per hour; the latter may be arrested safely barehanded, without a catcher's mitt, while I wouldn't recommend this method for the former (in fact, you'd need to stand pretty close to the pitcher in order to catch the slower ball before it fell to the ground).

Similarly, "angular momentum" is the rotational analog of "linear momentum." The rotation may be about an axis passing through the object, as is seen in a spinning top, or about a distant axis, exemplified by the moon orbiting the Earth. In quantum physics, the spin of electrons or protons resembles a top or a ballerina more than it does an orbiting satellite. Moreover, the spinning of the particles within an atom is not arbitrary but *must* correspond to particular values of angular momentum. This is like saying that the linear momentum of a car can have two values, moving forward or backward at multiples of a given speed, such as 10 miles per hour. So the car could go 30 miles per hour forward or 30 miles per hour backward, but not, say, 13 miles per hour in either direction.

It turns out, based on experimental observation, that certain fundamental particles in the universe have an internal angular momentum that has a value of either 0 (a very special case) or Planck's constant, h, divided by 2π. Photons, for example, have an intrinsic angular momentum of $h/2\pi$. Other fundamental particles, such as electrons, protons, and neutrons, can have an internal angular momentum of exactly one-half of this value of Planck's constant, h, divided by 2π, that is, $(1/2) \times (h/2\pi)$. That's it. Whether an object has an internal angular momentum that is either an integer multiplied by $h/2\pi$ or a half-integer multiplied by $h/2\pi$ will have a *profound* effect on how it interacts with other identical particles.

As 2π is just a number, if $h/2\pi$ is a measure of angular momentum then Planck's constant, h, is a unit of angular momentum. When Planck introduced the constant h as a fudge to account for the spectrum of light emitted by hot, glowing objects, he had hit upon a fundamental constant of the universe. There is a set of basic numbers that one must specify when setting up a universe, such as the mass of the electron and the speed of light. Things would look very different if the speed of light, for example, was a value a person could achieve while riding a bicycle, such as 15 miles per hour. One would then have an intimate, firsthand intuition about the consequences of the Special Theory of Relativity. Similarly, if Planck's constant were a much larger number, we would have to deal with quantum phenomena in our daily lives.

In Isaac Asimov's novel *Fantastic Voyage II: Destination Brain*, a team of scientists is reduced in size, smaller than a single cell, in order to travel within the body of an injured scientist (who has figured out a way to make miniaturization energy efficient!) and perform an operation. Asimov proposes that the mechanism underlying this shrinking process involves creating a field that reduces the magnitude of Planck's constant. Considering an atom to be a sphere, Bohr calculated its radius to be a few times r_o, where $r_o = h/[(2\pi)m_e c\, \alpha]$ and m_e is the mass of the electron, c is the speed of light, and α is termed "the fine structure constant" that involves another collection of fundamental constants (such as h, c, and the charge of the electron). If one could tune Planck's constant at will, making it larger or smaller, then one could enlarge or shrink any object by changing the fundamental size of its atoms.* The fact that we cannot do this in reality reflects the fact that fundamental constants are just that—constant and unchanging.

It was unnerving to physicists when Albert Einstein suggested in 1905 that there was no velocity faster than light speed, but the universe and its laws indeed ensure that nothing can move faster than the speed of light in a vacuum. Apparently, with the discovery that subatomic particles have internal angular momentums whose values are multiples of either $h/2\pi$ or $1/2$ of $h/2\pi$ but not any other values, the universe also cares about rotation.

The electron is the basic unit of negative charge, while the proton has an equal charge, but of an opposite sign (by convention the proton's charge is termed "positive" while the electron's is "negative"). It has been known since the 1820s that moving electric charges, that is, an electrical current, create a magnetic field. This is the basic physical principle underlying electromagnets and motors. If an electrically charged sphere rotates about a line passing through its center, like a wheel about its center spoke, then there are certainly electrical charges in motion, and these currents will generate a magnetic field. If there is an intrinsic angular momentum, it shouldn't be surprising that as a consequence of this rotation every electron and proton has its own internal magnetic field. In fact, the quantized internal angular momentum aspect was proposed to account for the experimentally observed internal magnetic fields inside atoms. That is, the observation of the magnetic field came first, and later, in an attempt to account for it, the argument about intrinsic angular momentum was put forward.

Does the experimentally observed magnetic field of electrons and protons actually arise from the spinning rotation of elementary particles? Technically, the answer is no. The simplest reason why

---

* When I described Asimov's suggestion in my 2005 book *The Physics of Superheroes*, it caught the attention of Tony Stark! In Marvel Comics's *Civil War Files*, consisting of background notes dictated by Iron Man's alter ego, under an entry for "Goliath," one finds: "My first introduction to Bill Foster (Goliath, Giant-Man, Black Goliath) was his remarkable paper with Jim Kakalios on extra-dimensional manipulations of Planck's constant, and I quickly had them hired by Stark Enterprises." Upon reading this, I realized that I needed to update my resumé!

not is that neutrons, the other fundamental particle found within atomic nuclei, which have nearly the same mass as protons but are electrically uncharged, *also* possesses an internal magnetic field! If the magnetic field of the proton arose from the fact that as a charged object, its rotation could be described as a series of electrical current loops, each of which generates a magnetic field, then the rotation of an electrically uncharged object should not generate a magnetic field.*

Moreover, even if we did not know that neutrons existed, we still could not explain the magnetic field of electrons as arising from their rotation about an axis passing through their center. The magnitude of the measured magnetic field of the electron is such that it would require that these particles spin at a rate so fast that points on their surface would be moving faster than the speed of light!

What experimental question does the proposal of intrinsic angular momentum (that is, spin) answer? In 1922 Otto Stern and Walther Gerlach passed beams of atoms through special magnets, looking for interactions between their laboratory magnet and the internal magnetic field of the atom. They were trying to probe the magnetic field that would arise from the electron's orbital motion about the nucleus. For atomic systems where they did not expect to see any orbital motion, they nevertheless observed an intrinsic magnetic field of elementary particles and, moreover, that this came in two values. It was if the electron had a built-in magnetic field with a north pole and a south pole that was allowed to point in only two directions, relative to the magnets used in Stern and Gerlach's experiment. The electron either pointed in the same direction as the external magnet, so that its north pole faced the south pole of the lab magnet, or exactly oppositely aligned, so that the electron's north pole faced the lab magnet's north pole.

While Stern and Gerlach's experiment clearly suggested that electrons possessed an intrinsic magnetic field, spin was actually first proposed to account for features in the absorption and emis-

---

* This argument still holds, even with the recognition that neutrons (and protons) are themselves composed of electrically charged quarks. As explained, the quarks would have to be rotating faster than light speed to account for the observed magnetic field of these composite particles.

sion of light by certain elements that indicated that there had to be some internal magnetism inside the atom. A variety of careful experiments confirmed that this magnetic field did not arise from the electrons orbiting around the positively charged nucleus, but was somehow coming from the electrons themselves.

So, why do we say that electrons, protons, and neutrons have spin that is associated with their internal magnetic fields? The origins of this phrase go back to, shall we say, a "youthful indiscretion." Two Dutch graduate students in Leiden, Samuel Goudsmit and George Uhlenbeck, wrote a paper in 1925 suggesting that an internal rotation of the charged electron generated a magnetic field necessary to account for the atomic light-emission-spectra anomalies. They showed their paper to their physics adviser, Paul Ehrenfest, who pointed out various problems with the electron literally spinning about an internal axis. Hendrik Lorentz, Ehrenfest's predecessor, soon calculated, as the students' argument required the electron to rotate faster than light speed, that, thanks to $E = mc^2$, this would make the electron heavier than the proton. (Neutrons, which would have indicated to them immediately that the observed magnetic field could not result from a spinning charged particle, had not yet been discovered.) Defeated, Goudsmit and Uhlenbeck intended to drop the whole matter. They were surprised when Ehrenfest told them that that he had already submitted their paper for publication. He consoled them, indicating that he had recognized their error but thought that there was merit in their suggestion, and argued that they were "young enough to be able to afford a stupidity."

On one level it is unfortunate that Goudsmit and Uhlenbeck employed the term "spin" for the intrinsic angular momentum and magnetic field possessed by subatomic particles. The term is so evocative (and the fact that an electron, for example, can have an intrinsic angular momentum of either $+(1/2)h/2\pi$ or $-(1/2)h/2\pi$, but no other values, makes it easy to think of in terms of "clockwise" or "counterclockwise" rotation) that it is difficult to remember that the electron is not actually spinning like a top. The intrinsic angular momentum of the electron is properly accounted for in the Dirac equation, a fully relativistic version of quantum theory. Solving the Dirac equation, one finds that the electron is characterized by an extra "quantum number" that corresponds to an internal angular momentum of $(1/2)h/2\pi$ and a magnetic field of magnitude exactly as observed. In a sense it is an intrinsic feature of the electron, just like its mass and its electric charge.* Goudsmit and Uhlenbeck managed to get the right answer for the wrong reasons. For this work they were awarded several elite prizes and medals. Publish in haste, celebrate at leisure.

I have promised that this internal angular momentum, possessed by electrons, protons, neutrons, and all other elementary particles, is the key to understanding the periodic table of the elements, chemistry, and solid-state physics. In Chapter 12 I will describe the Pauli exclusion principle, which states that when two electrons (or two protons or two neutrons) are so close to each other that their de Broglie waves overlap, they can both be in the same quantum state only if one electron has a spin of $+h/2\pi$ and the other has a spin of $-h/2\pi$. This has the consequence of "hiding," except in certain cases, the magnetism associated with the intrinsic angular momentum.

Electrons can spin either "clockwise" or "counterclockwise" (once an axis of rotation has been specified), which indicates that their intrinsic magnetic fields can point either "up" or "down." All magnets found in nature have both north and south poles. If we make a magnet in the shape of a cylinder, like a piece of chalk, then, as shown in Figure 10, there will be a magnetic field emanating from the north pole that will bend around and be drawn into the cylinder's south pole. The spatial variation of the magnetic field is the same as for an electric field created by two electrical charges, positive and negative, at either end of a cylinder (see Figure 10b). We call such an arrangement of electric charges a "dipole," and as the magnetic field distribution described earlier has the same spatial variation, we refer to it as a magnetic dipole. Inside an atom, the preferred configuration of the protons and neutrons in the nucleus, and electrons "orbiting" the nucleus, is such that they orient them-

---

* Many physicists, when pressed, would confess that the notion of the electron they carry around in their heads involves a particle with a large arrow protruding from it, pointing either "up" or "down," whenever "spin" comes up. Thus, if you find the image of an electron spinning like a top too compelling to give up—you're just thinking about it the way we professionals do!

selves so any pair of particles will have their magnetic fields cancel; thus, if the north pole of one magnet points "up," then the north pole of the second magnet will point "down."

The electric dipole field differs from a single positive or negative charge, which is called a "monopole," as shown in Figure 10a. We have never observed in the universe a single free magnetic pole, that is, just a north pole or a south pole, despite extensive investigations and theoretical suggestions that they should exist. They always come in pairs, forming a magnetic dipole. It must be said, though, that an unsuccessful search does not mean that they do not exist— simply that we haven't found them yet.

We need to figure out magnetism if we want to understand how hard drives work. Furthermore, without appreciating the role that spin plays, chemistry would be a mystery (unlike when most of us studied it in high school, when it was a hopeless mystery). Similarly, absent our understanding how the spin of electrons governs their interactions in metals, insulators, and semiconductors, there would be no transistor, and hence no computers, cell phones, MP3 players, or even television remote controls, and humanity would be reduced to a brutal state that would test the imagination of the writers of the most dystopian science fiction stories.

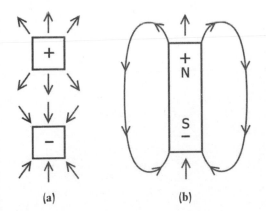

(a)   (b)

Figure 10: Sketch of the electric field from an isolated positive and negative charge (a) and from the two charges forming a dipole pair (b). The same field lines are found for a magnetic dipole, where the north pole plays the role of the positive charge, and the south pole acts like a negative charge.

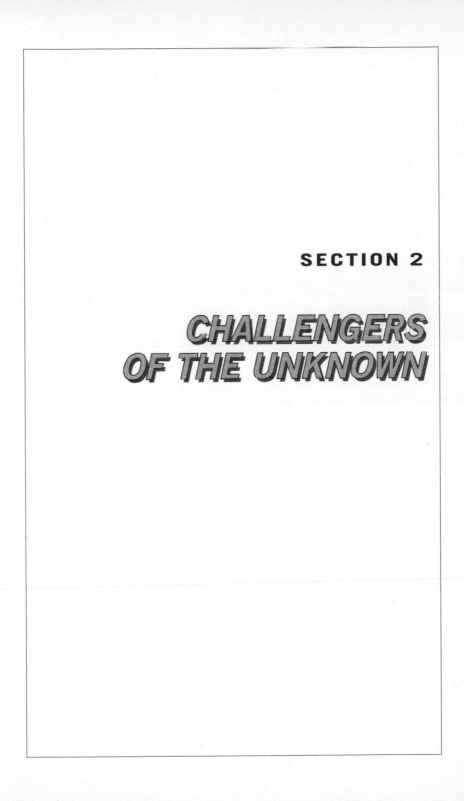

SECTION 2

*CHALLENGERS
OF THE UNKNOWN*

CHAPTER FIVE

Wave Functions All the Way Down

In 1958, Jonathan Osterman (Ph.D. in atomic physics, Princeton University) began his postdoctoral research position at the Gila Flats Research Facility in the Arizona desert. There
he participated in experiments probing the nature of the "intrinsic
field." This is the collection of forces responsible for holding all
matter together, aside from gravity. This would include electromagnetism, to account for the negatively charged electrons attracted to
the positively charged protons in the atom's nucleus. Electricity is
much stronger than gravity, so much stronger, in fact, that the
electrostatic attraction between the electrons and the protons in
an atom's nucleus is more than one hundred trillion trillion trillion
times stronger than their gravitational attraction to one another.
Consequently we can indeed neglect gravity's contribution to the
intrinsic field holding matter together.

In addition to electromagnetism, the intrinsic field must be
comprised of additional forces that act on the protons and neutrons
within each atom's nucleus. While opposite electrical charges are
attracted to each other, similar electrical charges are pushed away.
The nucleus contains positively charged protons that are electrostatically repelled from one another and electrically uncharged
neutrons that are immune to the electrostatic force. The closer two
charges are, the greater the electrical force between them. Given
that the protons within a nucleus are less than ten trillionths of a
centimeter from each other, the electrical repulsion between pro-

tons is very powerful, and any force capable of overcoming this must be very strong—so strong, in fact, that physicists named it the strong force,* and it is also therefore a component of the intrinsic field binding matter together. The strong force was originally believed to operate between protons and neutrons within the nucleus, holding them together. With theoretical and experimental investigations indicating that each of these nuclear particles is composed of quarks (which are in turn electrically charged), the strong force is now identified as the force that holds the quarks together and bleeds outs to neighboring particles within the nucleus. The experimental evidence for this force is indirect, as it turns out to be very difficult to slice protons and neutrons open and probe the quarks directly. But clearly there must be an attractive force operating within the nucleus that is able to overcome the electrostatic repulsion between protons at very close quarters.

In addition, physicists were unable to explain why certain elements' nuclei decay and emit high-energy electrons despite the binding glue of the strong force. The neutron, discovered in 1932 by James Chadwick, was found to be unstable. A neutron out in free space has a half-life of roughly fifteen minutes. That is, in a quarter of an hour, a neutron outside of a nucleus has a 50 percent chance of decaying into a proton, an electron, and a neutrino (technically an antineutrino). As the neutron is uncharged, this has nothing to do with electromagnetism, and as it is outside of a nucleus, the strong force does not apply. This decay is also found to occur for neutrons inside certain nuclei.

There must be some other type of force that can turn a neutron into a proton. This additional force can't be stronger than the strong force (or else nuclei would not hold together at all), but it appears to be not simply electromagnetism. This somewhat weaker force is termed, creatively enough, the weak force, and it is the third component of the intrinsic field. The strong force within the nucleus is roughly a hundred times stronger than electromagnetism

---

* Here physicists seem to have anticipated superhero comic books. In *X-Factor* # 72, when Guido, a superstrong member of a team of superpowered mutants, realized that nearly all other such teams have at least one member whose superpower involves superstrength, he adopted the code name Strong Guy.

(which is why nuclei with 90 to 100 protons, such as uranium and plutonium, are stable), while the weak force is one hundred billion times weaker than electromagnetism.

Physicists at the Gila Flats facility in the late 1950s, attempting to investigate the nature of the intrinsic field that governed the inner working of the atom, studied what happened when the intrinsic field was removed. This was essentially a variation on the traditional experimental technique to study subatomic matter—smash the atoms together. Actually these "atom smashers" study collisions not of atoms but of their building blocks, such as protons and electrons. Particle accelerators push protons* around a ring at velocities very near the speed of light. When they collide with a fixed target or with another beam of protons traveling in the opposite direction, the large energy of the collision can enable the generation of other, exotic particles, providing a practical application of Einstein's relation $E = mc^2$. It is through such violent reactions that the structure of matter has been elucidated.

The physicists attempting to explore the intrinsic field took essentially a subtler, though no less destructive, approach. To isolate and remove the intrinsic field, they created *another* intrinsic field that was completely "out of phase" with the original field. Here they applied the same principle of interference, illustrated in Figure 4b, that certain sound-elimination devices employ—creating sound waves that are 180 degrees out of phase with the ambient sound such that the two wave fronts completely cancel each other, resulting in the removal of the original sound. Similarly, in order to cancel the intrinsic field of matter, one must identify the frequency and phase of the field and create an identical field with the same amplitude but exactly out of phase with the original. Thus, it takes an intrinsic field generator to perform an intrinsic field subtraction.

Without the electromagnetic, strong, or weak forces, there is nothing to hold atoms or nuclei together, and all matter would rapidly and violently be torn apart. Unfortunately for Dr. Osterman, just such a fate befell him when he was accidentally locked

---

* Some accelerators use electrons or atomic nuclei instead of protons—and some accelerate particles in a straight line—but the idea is the same.

in the intrinsic field chamber during one such test run in 1959. Osterman, along with concrete block no. 15, which was the intended target of that day's intrinsic field removal experiment, was completely dematerialized once his strong, weak, and electromagnetic forces were negated. Through a process that remains poorly understood to this day, Osterman was able to re-create himself, atom by atom, cell by cell, to become the superpowered being known as Dr. Manhattan.

Jon Osterman (also known as Dr. Manhattan) was created by Alan Moore and Dave Gibbons in the graphic novel *Watchmen* (originally published in 1986–87 as a twelve-issue miniseries). Figure 11 shows the scene from *Watchmen* when Osterman first manifests in his post–intrinsic field extraction form. One of the most

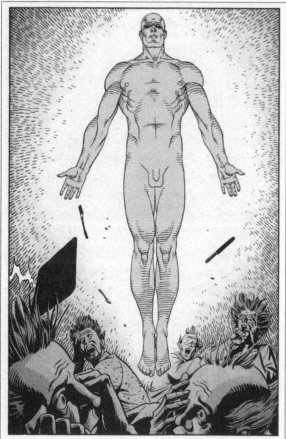

Figure 11: *Physicist Jon Osterman when he first reassembled himself following the removal of his "intrinsic field" in the graphic novel* Watchmen.

striking aspects of the reborn Dr. Manhattan is that his skin is bright blue, while Osterman had been a fairly typical Caucasian male. (Another feature of Dr. Manhattan that is hard to miss is that when he first successfully reintegrated his corporeal existence over the lunch tables in the Gila Flats cafeteria, he was completely naked. This aspect of the story turns out to be accurate—we physicists are extremely secure in our sexuality!)

Purely from a practical standpoint, the molecular chemical bonds that link the trillions and trillions of atoms in a person represent a vast amount of stored electrostatic energy. In order to cancel out, the "intrinsic field" for a person would require the energy of a nuclear power plant at full capacity. Even assuming that we could turn off the strong, weak, and electromagnetic forces, there is no way that any poor scientist subjected to such a procedure would be able to reconstitute himself following this ordeal. But, in considering the array of abilities that the character Dr. Manhattan displays in the pages of *Watchmen*, there does appear to be some interesting physics at play.

In addition to his now bright blue appearance, Osterman gained the ability to alter his size at will, to teleport himself and others instantly from one location to another, and to be aware of the future, that is, to experience time—past, present, and future—simultaneously. That is to say, following the removal of his intrinsic field and subsequent rebirth as Dr. Manhattan, Jon Osterman appears to have gained independent control over his quantum mechanical wave function.

His quantum mechanical what? We have now reached the point in our amazing story where we consider what a wave function is, and why gaining control over it would be akin to possessing superpowers. Here's the short answer: The quantum mechanical aspect of any object is reflected in its wave function. By performing simple mathematical operations on the wave function, one can calculate the probability density (that is, the probability per unit volume) of finding the object, whether it is an electron, an atom, or a large blue, naked physicist, at any point in space and time. If you could indeed alter your wave function at will, you would gain the ability to instantly appear at some distant location, without ever technically traveling between your initial and final points; you could change your size (from either very large to tiny); you could diffract into multiple versions of yourself; and you would be cognizant of your future evolution. And you'd likely give off a blue glow, though as we'll see later, that is more a consequence of leaking high-energy electrons—a side effect of rebuilding yourself at the atomic level.

*　*　*

While Jon Osterman is not a real person, nor is there a wave associated with an "intrinsic field," nor any such thing as an "intrinsic field," for that matter, the rest of the preceding discussion about the fundamental forces of nature (that is, electromagnetism and strong and weak nuclear forces) was correct. The experiments described in Section 1 demonstrated that there *is* a wave associated with the motion of electrons and atoms, and in fact with the motion of any and all matter. The concept of a wave function, introduced by Erwin Schrödinger in his "matter-wave equation," is the key to understanding all atomic and molecular physics. It might as well be called the "intrinsic field" for the central role it plays in the understanding of chemical bonding, by which all matter is held together.

At the start of the twentieth century, physicists debated whether the electrical charges in an atom were spread out throughout the atom, uniformly distributed in space, or existed as concentrated pointlike negatively charged electrons and positively charged protons. A series of experiments by Ernest Rutherford, Ernest Marsden, and Hans Geiger convinced scientists by 1911 that atoms were comprised of massive, positively charged protons in a physically very small nucleus, toward which the lighter, negatively charged electrons are electrically attracted. This is the familiar "solar system" picture of the atom that you might remember from grade school. But an electrically charged solar-system atom has a big problem in stability.

The Earth is pulled toward the sun by gravity, so what keeps our planet from falling into the sun? It turns out that this is a common misconception—the Earth is falling into the sun all the time! Don't panic—we're not on a death spiral to a fiery end. The Earth is moving at a high velocity at nearly a right angle to an imaginary line connecting us to the sun. The gravitational pull toward the sun deflects the Earth away from its straight-line path (an object in motion will remain in motion—unless acted upon by an outside force, such as gravity from the sun). The combination of the acceleration toward the sun, and the motion at a right angle away from the sun, results in a circular trajectory (the actual orbit is an ellipse—a distorted circle). The stable orbits of the planets in our solar system are possible only through the continual falling toward the sun. The Earth maintains its orbital motion, as there is nothing to slow it down (collisions with space particles provide a very small frictional drag that we can neglect), while the conditions that led to the Earth's original velocity about the sun are, as they say, a subject of current research.

As the force of electrical attraction between the electrons and protons in an atom is mathematically similar to the attractive force of gravity, a completely analogous argument suggests that the electrons should move in circular or elliptical orbits about the nucleus, not unlike the way the planets orbit the sun in our solar system. The main difference is that planets do not emit energy as they orbit the sun, but orbiting electrons do lose energy—in fact, quantum mechanics was developed in part to explain why all atoms *don't* suffer a death spiral to oblivion.

It was known from the days of the American Civil War that whenever an electric charge changes its direction, as in an elliptical orbit about an atomic nucleus, it emits electromagnetic radiation—that is, light. Since light carries energy, the electrons should lose energy as they orbit, and the slower they move, the less they are able to resist the attractive pull of the positively charged protons in the nucleus. In a short time (actually, less than a trillionth of a second), they should spiral into the nucleus. However, atoms form chemical bonds with other atoms, by which materials such as table salt, sand, and DNA are possible. The chemical bonds holding molecules and solids together involve interactions of the orbiting electrons among neighboring atoms, which would not be possible if the electrons were sitting on the nuclei. Something in the picture had to be wrong. If accelerated electric charges did not emit electromagnetic radiation, then radio and TV would not be possible.*

An important step in reconciling this puzzle was Niels Bohr's

---

* In such a world, to quote Krusty the Clown, "the living would envy the dead."

suggestion in 1913 that the electrons in an atom can assume only particular trajectories about the nucleus. That is, only certain planetary-like orbits are allowed. Electrons can jump from one orbit to another, but they may not follow any arbitrary path around the nucleus. An analogy: The city of Minneapolis in Minnesota contains a series of lakes that can be circumnavigated by paved paths. There are several paths encircling each lake, one intended for pedestrians, another for bicyclists, and a third for automobiles, and each pathway is separated from the others by a grassy median. Bohr's electronic orbits were analogous to these pathways, where electrons were free to travel but were forbidden from walking on the grass, as shown in Figure 12. The closer the orbit was to the nucleus, the more tightly bound the electron would be, so that more energy would have to be supplied to remove an inner-orbit electron than would be required to remove one from an outer ring. The electron could jump from an outer pathway to an inner loop, with the emission of the appropriate amount of energy, say by emitting light. Alternatively, by absorbing just the right amount of energy, it could be promoted from an inner orbit to a higher-energy, outer orbit (provided that the path had an open, available space for the electron). Bohr proposed that, for some reason that he could not explain, the electron would not emit light during its orbit, despite the requirements of Maxwell's equation for a charge that is constantly changing direction, but rather give off light only when moving from one orbit to another.

Bohr's proposal that only certain discrete orbits were possible was an attempt to account for the spectrum of light emitted by different atoms. Why are neon signs red, while the light from sodium lamps has a yellow tint? Neon, sodium, and in fact all atoms have the color they do because the electrons in these atoms have a predominant absorption (and emission) at only specific colors of light. The chemical differences between neon, with ten electrons (and ten protons in its nucleus), and sodium, with eleven electrons (balanced by eleven nuclear protons), leads to the separation between the relevant orbits corresponding to light in the red and yellow portions of the electromagnetic spectrum, respectively.

One can test this quantum principle tonight at dinner. Sprinkle a little bit of salt into the table candle (not so much that you

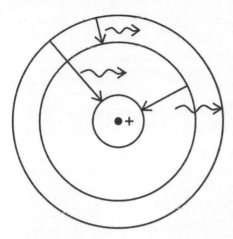

Figure 12: *Sketch of Bohr's proposed discrete electron orbits about a positively charged nucleus. Only certain trajectories are allowed, and an electron has a different energy depending on which orbital path it is on. The electron emits or absorbs light only when moving from one orbit to another.*

smother the flame) and you will see a distinct yellow tinge to the candle's light. The energetic atoms in the flame will excite electronic transitions in the sodium in the salt crystals, with electrons moving from one quantum orbit to another, and when the electrons return to their original orbits, they give off yellow light. In fact, independent of sodium, the light we see from a candle results from the electrons in the hot gas atoms near the wick being excited into high-energy orbits. When the electrons return to their lower energy states, they emit photons—which is the source of the light in a flame. We cannot have a complete understanding of fire—*the oldest* technology—without an understanding of quantum mechanics.

Each element has its own unique spectrum of absorption (or emission) lines, specific wavelengths of light that correspond to allowed transitions from one electron orbit to another. Just as in the case of fingerprints or snowflakes, no two elements have exactly the same spectrum of absorption lines. By measuring the different wavelengths of the light emitted by an atom, one can identify the element or molecule. In fact, this is how the element helium, the second most common element in the universe (after hydrogen), was discovered. When absorption lines from light from the sun were studied, the line spectra for hydrogen were observed, but there was another series of lines that did not correspond to any element known on Earth. This newly detected element was named helium after the Greek god of the sun, Helios.

If you want to see what a line spectrum for a previously un-

Figure 13: *Accurate illustration of an absorption spectrum from a 1955 DC Comics science fiction comic book. The element in question only absorbs light at very particular wavelengths, providing a unique signature as to its elemental composition. This line spectrum is a hallmark of the quantum nature of atoms.*

known element would look like, consider the November 1955 issue of *Strange Adventures*, a science fiction anthology comic book published by National Allied Periodicals, home of Superman and Batman. As shown in Figure 13, scientist Ken Warren uncovers the existence of a "radioactive metallic element hitherto* undiscovered in the entire solar system." In order to trace the source of this new metallic element, Dr. Warren uses a scintillometer. A caption box in the story informs the readers that this is a device capable of detecting "even the smallest amount of radioactivity." While nowadays a scintillometer refers to a device that detects small variations in the optical properties of the atmosphere, back in 1955 this was indeed the term used to measure the presence of ionizing radiation. Dr. Warren discovers that the radioactive element was brought to Earth by an alien spacecraft. This in turn

---

* Comic books back in the 1950s, at the height of Fredric Werthem's *Seduction of the Innocent* scare, may indeed have been corrupting young readers' minds (after all, isn't that what literature is *supposed* to do?), but one could hardly complain that they weren't improving readers' vocabulary or reading comprehension.

leads to his capture by two would-be invaders who intend to set the entire planet aflame. Apparently if Earth could be converted into a flaming sun, then the temperature on the moon would rise to a point that would accommodate the aliens' physiology. The fact that the sun is not actually a large planet that has been set on fire seems to have escaped these aliens, who have nevertheless managed to master interstellar flight, but it turns out to be a moot point, as Dr. Warren and his chemist colleague Hank Forrest are able to trick the aliens into abandoning their plan and leaving our solar system.

In a sense, Bohr's proposal that atomic line spectra arise from energy emission or absorption from electrons residing in discrete orbits managed only to reframe the mystery of atomic spectra. Now, instead of wondering why atoms absorbed or emitted light at only particular wavelengths, one could ask why only certain electronic orbits were possible about the atomic nucleus. Prince de Broglie's matter-waves provided an answer.

Schrödinger was one of the first physicists to recognize that de Broglie's matter-wave hypothesis could neatly solve this riddle of why the electrons did not collapse into the nucleus. For if there is indeed a wave associated with the motion of the electron, as demonstrated by Thompson's observations reported in the pages of *Science Wonder Stories* (and the experiments shown in Figure 7), then this wave has to be a "standing wave."

A guitar string, clamped at both ends, as shown in Figure 14, cannot oscillate with any wavelength; only those waves are possible for which the amplitude is zero at the two fixed ends. This commonsense constraint leads to a very restricted set of possible waves that the string can support—which is why when plucked the

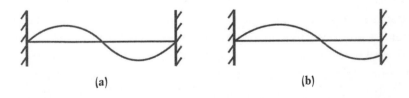

(a)   (b)

Figure 14: *Sketch of an allowed standing wave for a vibrating string clamped at both ends (a) and a wave that is not possible (b).*

guitar string vibrates only at certain frequencies. Wavelengths as illustrated in Figure 14b are simply not possible. This is the whole point of a musical instrument, after all, as a string that vibrated at *all* frequencies would be pretty useless for constructing harmonies. The possible waves of a clamped string, constrained by the fixed ends and unable to travel down the length of the string, are called "standing waves."

Say there is a wave associated with us as we walk along the pedestrian pathway circling Lake Harriet in Minneapolis (and we do in fact have a "matter-wave" moving along with us, but of such a small wavelength that it is undetectable). Once we have returned to our starting point, our wave must join up smoothly and perfectly with the wave started when we left for our stroll. If the wave were at its highest point, a crest, when we began our walk, then as we walked around the lake, the wave would oscillate down to a valley, back up to a crest, and so on. Once we are back at our starting point, the wave must again be at a crest. This simple, commonsense aspect of waves (what would it look like if the cycle of the wave were such that our wave was at a valley when we returned to our starting point, where the original wave was a peak?) leads immediately to the consequence that only certain pathways—those that correspond to different wavelengths that start and end correctly— around the lake are possible.

Similarly, a wave associated with an electron in a closed orbit, returning to the same point after a full rotation, can assume only certain wavelengths. Once an orbit has been completed, the wave must be at the same point as when it left. This "single-valued" constraint, that at any point the wave can have only one amplitude (that is, it can't be a peak and a valley at the same time), restricts the infinite range of possible wavelengths to a very small set of allowed orbits. Just as the guitar string has a lowest pitch—it is impossible to excite a wave on the clamped string lower than its fundamental oscillation—there is a lowest standing-wave electron orbit that can be constructed around the nucleus. Thus, electrons do not continuously lose energy and spiral into the nucleus with ever-decreasing radii, as there is no way for the apparently very real matter-wave to form standing waves with a lower wavelength than the lowest possible orbit. The de Broglie hypothesis of matter-waves saves the stability of atoms and also accounts in a natural way for the line spectra (as shown in Figure 12) of atoms.

Following a seminar presentation by Schrödinger of the matter-wave model of electrons in an atom, Pieter Debye, a senior physicist in the audience, pointed out that if electrons had waves, there should be a corresponding wave equation to describe them (just as Maxwell had found a wave equation for electric and magnetic fields that turned out to accurately describe the properties of electromagnetic waves, that is, light) and challenged Schrödinger to find it. Taking up this assignment, Schrödinger accomplished the deed in just six months. A consequence of Schrödinger's work was the realization that the electrons could not really be considered as mini-planets in a nuclear solar system. The resulting equation, which now bears his name, would garner him a Nobel Prize, change the future, and lead to philosophical arguments over the nature of measurements of quantum systems that so disturbed Schrödinger that he would claim that he was sorry he brought the whole thing up.

CHAPTER SIX

The Equation That Made the Future!

The Schrödinger equation plays the same role in atomic physics that Newton's laws of motion play in the mechanics of everyday objects. Those with long memories may recall that back in the seventeenth century it was well-known that the application of external forces was required to change the motion of objects. What was lacking was a method by which one could calculate exactly *how* the motion of any given object would change as a consequence of the pushes and pulls of external forces. Newton found a remarkably elegant expression (the net force is equal to the mass multiplied by the acceleration, or $F = m \times a$) that despite its surface simplicity could account for a wide range of complex motions.

As any student in an introductory physics class can tell you, it is not enough to identify the forces acting in a situation—one must be able to show how these forces will lead to a change in the object's motion. Armed with Newton's law, one can determine the trajectory of skiers and boaters, of runners and automobiles, of rockets fired at the moon, and of the moon itself (not to mention falling apples). We know that Newton's laws of motion are correct, for comparisons of the theoretically calculated motion agree exactly with what is experimentally observed.

Similarly, in the quantum realm, it is not enough to say that there is a wave associated with the motion of all matter whose wavelength is inversely proportional to its momentum. One also needs a process—an equation—by which, if one knows the external forces acting on the object, the resulting behavior of its "matter-wave" can be determined.

Consider a simple hydrogen atom with one proton in its nucleus and one electron attracted to the proton electrostatically, the symbol that Dr. Manhattan etched into his forehead in the *Watchmen* comic and film. Given that we know the nature of the electrostatic attractive force between the negatively charged electron and the positively charged nucleus and that there is a wave associated with the motion of the electron in an atom, what we would like is a "matter-wave equation" that enables us to calculate the properties of the electron in the atom. These calculated properties, such as the average diameter of the atom, or the electron's average momentum, could then be compared to experimental measurements, in order to test the correctness of the matter-wave approach. Any equation that does not yield testable results is useless from a physics point of view.

Schrödinger found just such an equation, though the procedure for calculating the wave function involves a formula *much* more complicated than Newton's $F = m \times a$. In fact, one often requires a computer in order to solve Schrödinger's equation, except in a handful of simple cases. But the fact that the calculations can be difficult does not invalidate the Schrödinger approach. Applying Newton's law of motion, $F = m \times a$, to the motion of the more than trillion trillion air molecules in a room exceeds the calculating capability of the largest supercomputer. Even though we can't do the sums, we know that in principle they could be done. The importance of Schrödinger's equation is that the process by which one calculates an object's wave function, given the relevant forces acting on it, is known. The math might be hard, but the path is clear.

We will not go into the process, involving mathematical trial and error, physical intuition, and a creative use of the principle of conservation of energy that enabled Schrödinger to develop his equation. It is, indeed, a fascinating story, filled with false starts, suspense, and plenty of sex!* We are interested in how quantum

---

* I'm not kidding! While other pioneers of quantum theory such as Werner Heisenberg and Paul Dirac required extreme solitude in order to develop their theories, Schrödinger needed a more . . . stimulating environment. Historians of science debate to this day the identity of the woman (definitely not his wife!) who kept him company over a long Christmas holiday break, at a friend's chalet in the Swiss Alps, when he successfully

mechanics brought about our futuristic lifestyle of light-emitting diodes, laptop computers, cell phones, and remote controls. Consequently, for our purposes, it is not important how the Schrödinger equation was developed, nor will we try to solve it, even for the simplest case of a single electron moving in a straight line in empty space (and it doesn't get more plain vanilla than that). Schrödinger's equation involves the rates of change of the wave function in both space and time; consequently it can't be solved without calculus. In addition, it involves imaginary numbers (the term mathematicians use when referring to the square root of negative numbers), and thus calls upon considerable imagination to interpret. What we *will* do is discuss the significance of the equation and point out how various solutions lead, in some cases, to semiconductor transistors and, in other situations, to semiconductor lasers.

Enough with the tease: The Schrödinger equation, as it is now universally known, is most often expressed mathematically as follows:

$$-\hbar^2/2m \; \partial^2\Psi/\partial x^2 + V(x, t) \; \Psi = i \, \hbar \, \partial\Psi/\partial t$$

where $\hbar$ represents our old friend, Planck's constant (the angled bar through the vertical line in the letter h is a mathematical shorthand that indicates that the value of Planck's constant here should be divided by 2π, that is, $\hbar = h/2\pi$); m is the mass of the object (typically this would be the electron's mass) whose wave function Ψ (pronounced "sigh") we are interested in determining; V is a mathematical expression that reflects the external forces acting on the object; and i is the square root of -1.* The rate of change of Ψ in time is represented by $\partial\Psi/\partial t$, while the rate of change of the rate

---

constructed the matter-wave equation that now bears his name. As I tell my students of modern physics, if you are going to learn quantum mechanics, you must remember one key principle: Don't be a playa' hata'!

* Since the mathematical convention is that a positive number multiplied by a negative number yields a negative number, and the product of two negative numbers is a positive number, there is no real number that, when multiplied by itself, would yield a negative number. We can certainly imagine such numbers, but they are not in the set of real numbers we deal with. They are hence termed "imaginary" and are defined to be represented by the lower case letter i, so $i \times i = -1$.

of change of Ψ in space is represented by $\partial^2\Psi/\partial x^2$. As messy and complicated as this equation may seem, I have taken it easy and written this formula in its simpler, one-dimensional form, where the electron can move only along a straight line. The version of this equation capable of describing excursions through three dimensions has a few extra terms, but we won't be solving that equation either.

The Schrödinger equation requires us to know the forces that act on the atomic electrons in order to figure out where the electrons are likely to be and what their energies are. The term labeled V in the Schrödinger equation represents the work done on the electron by external forces, which can change the energy of the atomic electron. For reasons that are not very important right now, V is referred to as the "potential" acting on the electron.*

In the most general case, these forces can change with distance and time. The fact that these forces, and hence the potential V, can be time t– and space x–dependent is reflected in the notation $V(x,t)$ in this equation, which implies that V can take on different values at different points in space x and at different times t. Sometimes the forces do not change with time, as in the electrical attraction between the negatively charged electron and the positively charged protons in the atomic nucleus. In this case we need only to know how the electrical force varies with the separation between the two charges, to determine the potential V at all points in space.

I've belabored the fact that V can take on different values depending on which point in space x we are, and at what time t we consider, because what is true of V is also true of the wave function Ψ. Mathematically, an expression that does not have one single value, but can take on many possible values depending on where you measure it or when you look, is called a "function." Anyone who has read a topographic map is familiar with the notion of functions, where different regions of the map, sometimes denoted by different colors, represent different heights above sea level. This is why Ψ is called a "wave *function*." It is the mathe-

---

* We use the letter *V* for the potential, as the letter *p* is reserved for mathematical descriptions of the *mo*-mentum! Sometimes it seems like we physicists deliberately make the equations harder than they need to be.

matical expression that tells me the value of the matter-wave depending on *where* the electron is (its location in space x) and *when* I measure it (at a time t). Though as we'll see in the next chapter with Heisenberg, where and when get a little fuzzy with quantum objects.

For an electron near a proton, such as a hydrogen atom, the only force on the electron is the electrostatic attraction. Since the nature of the electrical attraction does not change with time, the potential V will depend only on how far apart in space the two charges are from each other. One can then solve the Schrödinger equation to see what wave function Ψ. will be consistent with this particular V. The wave function Ψ will also be a mathematical function that will take on different values depending on the point in space. Now, here's the weird thing (among a large list of "weird things" in quantum physics). When Schrödinger first solved this equation for the hydrogen atom, he incorrectly interpreted what Ψ represented—in his own equation!

Schrödinger knew that Ψ itself could not be any physical quantity related to the electron inside the atom. This was because the mathematical function he obtained for Ψ involved the imaginary number i. Any measurement of a real, physical quantity must involve real numbers, and not the square root of –1. But there are mathematical procedures that enable one to get rid of the imaginary numbers in a mathematical function. Once we know the wave function Ψ, then if we square it—that is, multiply it by itself so $\Psi \times \Psi^* = \Psi^2$ (pronounced "sigh-squared")—we obtain a new mathematical function, termed, imaginatively enough, Ψ^2.** Why would we want to do that? What physical interpretation should we give to the mathematical function Ψ^2?

Schrödinger noticed that Ψ^2 for the full three-dimensional version of the matter-wave equation had the physical units of a number divided by a volume. The wave function Ψ itself has units of

---

** The notation Ψ^* is mathematical code that says to change all the terms with i in Ψ to $-i$. As $i \times i = -1$, $(-i) \times i = -1 \times (i \times i) = -1 \times -1 = +1$, and we will have a positive, real function in Ψ^2. Those who recall their algebra may note that Ψ is actually a "complex" number (Ψ = a + ib, where a and b are regular numbers), so that I should technically use the notation $|\Psi|^2$ instead of Ψ^2. We'll stick with Ψ^2, as for us the important point is that Ψ^2 is a real, positive number.

1/(square root (volume)). This is what motivated the consideration of Ψ^2 rather than Ψ—for while there are physically meaningful quantities that have the units of 1/volume, there is nothing that can be measured that has units of 1/(square root (volume)). He argued that if Ψ^2 were multiplied by the charge of the electron, then the result would indicate the charge per volume, also known as the charge density of the electron. Reasonable—but wrong. Ψ^2 does indeed have the form of a number density, but Schrödinger himself incorrectly identified the physical interpretation of solutions to his own equation.

Within the year of Schrödinger publishing his development of a matter-wave equation, Max Born argued that in fact Ψ^2 represented the "probability density" for the electron in the atom. That is, the function Ψ^2 tells us the probability per volume of finding the electron at any given point within the atom. Schrödinger thought this was nuts, but in fact Born's interpretation is accepted by all physicists as being correct.

What Schrödinger discovered was the quantum analog of Newton's force = (mass) × (acceleration). Newton said, in essence, you tell me the net forces acting on an object, and I can tell you where it will be (how the motion will change) at some other point in space and at some later time. Schrödinger's equation asserts that if you know the potential V (which can be found by identifying the forces acting on the electron), then I can tell you the probability per volume of finding the electron at some point in space and time, now and in the past and future.

Once I know how likely it is to find an electron throughout space, I can calculate its average distance from the proton in the nucleus, which we would call the size of the atom. The energy within the atom, its average momentum, and in fact anything I care to measure can be calculated using the Schrödinger equation. We know that the Schrödinger equation is a correct approach for determining the wave function Ψ for electrons in an atom, for it enables us to calculate properties of the atom that are in excellent agreement with experimental observations. This is the only true test of the theory and the only reason why we take seriously notions of matter-waves and wave functions.

Schrödinger first applied his equation to the simplest atom, hydrogen, with a nucleus of a single proton and only one orbiting electron. For the force acting on the electron he used the familiar law of electrostatic attraction, extensively confirmed as the correct description of the pull two opposite charges exert on each other. With no other assumptions or ad hoc guesses, his equation then yielded a series of possible solutions, that is, a set of different Ψ functions that corresponded to the electron having different probability densities. This is not unlike the set of different possible wavelengths that a plucked guitar string can assume. For each probability density there was a different average radius, and in turn a different energy. Obviously the solitary electron in the hydrogen atom could have only one energy value and only one average radius at any given time. What Schrödinger found was that there was a collection of possible energies that the electron could have, not unlike an arrangement of seats in a large lecture hall (shown schematically in Figure 15). There are some seats close to the front blackboard, some in the next row a little bit farther away, and so on all the way to seats so far from the front of the room that a student sitting here could easily sneak out of the class with very little effort.

An atom with only one electron is akin to a lecture hall with only one student attending. If there is a tilt to the hall, so that the front of the room is at a lower level than the rear (as in many audi-

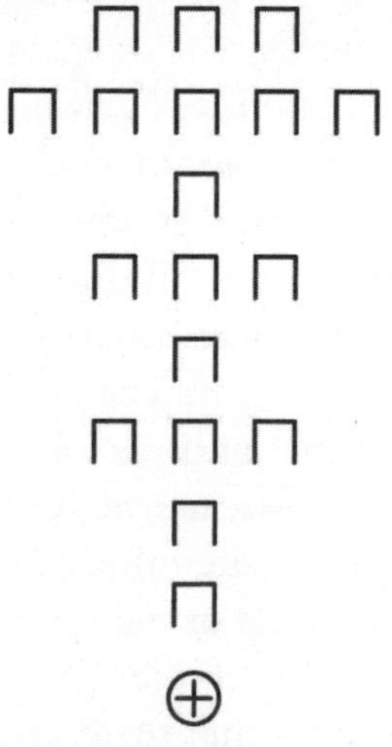

Figure 15: *Cartoon sketch of the possible quantum states, represented as seats in a lecture hall that an electron can occupy, as determined by the Schrödinger equation for a single electron atom. In this analogy the front of the lecture hall, at the bottom of the figure, is where the positively charged nucleus resides. Upon absorbing or releasing energy, the electron can move from one row to another.*

toriums), then the student would lower his or her energy by sitting in the front seat in the front row. This would be the lowest energy configuration for the student in the lecture hall, and similarly Schrödinger found that a single electron would have a probability density that corresponded to it having the lowest possible energy value. This configuration for an atom is termed the "ground state." If the electron received additional energy, say, by absorbing light or through a collision with another atom, it could move from its seat closest to the front of the room to one farther away. With enough energy it could be promoted out of the auditorium entirely, and in this case it would be a free electron, leaving behind a positively charged ionized atom (which for the simple case of hydrogen would be just a single proton).

Schrödinger's equation at once explained why atoms could absorb light only at specific wavelengths. There are only certain energies that correspond to valid solutions to the Schrödinger equation for an electron in an atom, pulled toward the positively charged nucleus by electrostatic attraction. Electrons ordinarily sit in the lowest energy level—the ground-state configuration. Upon absorbing energy from outside the atom, they can move from the row closest to the front of the hall to another row at a higher level, closer to the exit, provided the seats they move to are empty. Those who recall their high school chemistry may know that each "seat" can actually accommodate two electrons—this is discussed in some detail in Section 4. There is a precise energy difference between the row where the electron is sitting and the empty seat it will be promoted to. It must move from seat to seat and cannot reside between rows. Different atoms with different numbers of protons in their nucleus will have slightly different Ψ functions and different spacing between the rows of possible seats, similar to each string on a guitar having a different fundamental frequency as well as different overtones.

One intriguing consequence of the Schrödinger equation is that it explained that electrons in an atom could have only certain energy values, and that all other electronic energies were forbidden. Schrödinger's solution has nothing to do with circular or elliptical orbits, but rather with the different possible probability densities the electron can have. Another way of stating this is that the probability of an electron having a "forbidden energy" is zero and thus will never be observed to occur.

Only if the energy supplied to the atom, for example, in the form of light, is exactly equal to this difference will the electron be able to move to this empty seat. Too much energy or too little would promote the electron between rows. Since the electron cannot be at these energy values (the probability of it occurring is zero), the electron cannot absorb the light of these energies. There will thus be a series of wavelengths of light that any given atom can absorb, or emit, when the electron moves from one row to another; all other light is ignored by the atom.

Now for the cool part. When Schrödinger used the electrostatic attraction between a proton and an electron for a simple hydrogen atom, he found a set of possible wave functions that corresponded to different probability densities. When he made sure that Ψ was "normalized"—which is a fancy way of saying that if you add up the probability density of the electron over all space, the total has to be 100 percent—then his equation yielded a set of possible energy levels for the electron that *exactly* corresponded to the energy transitions experimentally observed in hydrogen. No assumptions about orbits, no ignoring the fact that electromagnetism requires the electron to radiate energy in a circular trajectory. All you have to tell the Schrödinger equation is the nature of the interaction between the negatively charged electron and the positively charged nucleus in the atom, and it automatically tells you the possible line spectra of light for emission or absorption.

Schrödinger's equation destroyed the notion of a well-defined elliptical trajectory for the electron and replaced it with a smoothly varying probability of finding the electron at some point in space. In essence, physics had come full circle. At the start of the twentieth century, scientists thought that the atom consisted of a jelly of positive charges, in which electrons sat like marshmallows in a Jell-O dessert. In this model the atom emitted or absorbed light at very particular wavelengths as these wavelengths corresponded to the fundamental frequencies of oscillation of the trapped electrons. Rutherford demonstrated that the atom actually had a small positively charged nucleus, around which the electrons orbited. But this model could not explain what kept the electron from spiraling into the nucleus, nor did it account for the absorption line spectra. With Schrödinger the small positively charged nucleus remained, but the negatively charged electrons were like jelly smeared over the atom, in the form of a "probability density." The source of the "uncertainty" in the electron's location was accounted for by Werner Heisenberg, independently of Schrödinger, who was mixing business with pleasure in the Swiss Alps.

CHAPTER SEVEN

The Uncertainty Principle Made Easy

Just prior to the introduction of the Schrödinger equation, Werner Heisenberg was developing an alternative approach to atomic physics. The wavelike nature of matter is also at the heart of the famous Heisenberg uncertainty principle. Heisenberg took a completely different tack to the question of how de Broglie's matter-waves could account for an atom's optical absorption line spectra. As we've discussed, the conventional view that the electron was executing classical orbits around the nucleus could not be reconciled with the absence of light that should be continuously emitted by such an electron. Heisenberg's struggle to envision what the electron was doing and where it was located within the atom finally convinced him to give up and not bother trying to figure out where the electron was. This turned out to be a winning strategy.

Physics is an experimental science, and the development of quantum mechanics was driven by a need to account for atomic measurements that were in conflict with what was expected from electromagnetic theory and thermodynamics. Heisenberg realized that any theory of the atom needed only to agree with measurements, rather than make predictions that could never be tested! Who cares where the electron "really was" inside the atom? Could you ever definitely measure its location to confirm or disprove this prediction? If not, then forget about it. What *could* be measured?

For one, the wavelengths of the light emitted from an atom in a line spectrum. Well then, let's construct a theory that describes the energy difference when an atom makes a transition from one state to another. Heisenberg's model for the atom consisted of a large array of numbers that characterized the different states the electron could be in, and rules that governed when the electron could go from one state to another. When compared to the observed spectra of light absorbed or emitted from a hydrogen atom, Heisenberg's approach agreed exactly with measurements.

In 1925 Heisenberg exiled himself to the German island of Helgoland in the southeast corner of the North Sea as he worked out the details of this approach. He had known for years that, in contrast to Schrödinger, he did his best work removed from any distractions. While the isolation suited his requirements for extended contemplation, his motivation for decamping to the island was a bit more mundane. Heisenberg suffered from severe allergies, and it was to escape the pollens of Göttingen that he traveled to the treeless Helgoland. The stark island's only recreational option consisted of mountains, of which the hiking enthusiast Heisenberg availed himself as he struggled with his theory.

On this island Heisenberg constructed arrays of values for the different states in which electrons in an atom could be found, and the rules for how they would make the transition between states. When he returned to Berlin and showed his preliminary efforts to Max Born, the older professor was initially confused. As he read and reread Heisenberg's work, trying to understand these transition rules, he had a strong sense of familiarity. Eventually Born and Pascual Jordan, another physicist at Göttingen, recognized that Heisenberg had independently developed a mathematical notation known as a "matrix" to describe his theory—a notation that had been invented a hundred years earlier by mathematicians interested in solving series of equations that had many unknown variables. This branch of mathematics is called linear algebra, or matrix algebra, and Heisenberg had unknowingly reinvented a wheel that had been rounded years earlier.

(This is always happening, by the way. More often than not we find in physics that the necessary mathematics to solve a particularly challenging problem has already been developed, frequently no less than a century earlier. The mathematicians are just doing what mathematicians do and are not trying to anticipate or solve any physics problems. There are two notable exceptions: Newton had to invent calculus in order to test his predictions of celestial motion against observations, and modern string theorists are inventing the necessary mathematics simultaneously with the physics.)

Heisenberg's approach using matrices is an alternative explanation for the behavior of electrons in atoms, for which he was awarded the Nobel Prize in Physics in 1932. Heisenberg's theory was published in 1925. Less than a year later Schrödinger introduced his matter-wave equation to account for the interactions of electrons inside the atom. The two descriptions of the quantum world do not appear to have anything in common, aside from the fact that they both accurately predict the observed optical line spectra for atoms. In 1926 Schrödinger was able to mathematically translate one approach into the other, demonstrating that the two descriptions are in fact equivalent. Both theories rely on de Broglie's matter-wave hypothesis, though neither takes the suggestion of elliptical electronic orbits literally. Both approaches make use of what we now would characterize as the electron's wave function and have prescriptions for how one can mathematically calculate the average momentum, the average position, and other properties of an electron in an atom. For our purposes, we do not have to go too deeply into either theory, as our goal is to understand how the concepts of quantum mechanics underlie such wonders of the modern age as the laser and magnetic resonance imaging.

*   *   *

There is one aspect of Heisenberg's theory that has generated way too much blather and misinformation to be ignored—that is, the famed uncertainty principle. If you'll bear with me, I'd like to take a brief detour to set the record straight about what the uncertainty principle does and does not mean. It's actually not too complicated, that is, once one accepts that there is a wavelike nature to matter.

The uncertainty principle posits a relationship between the uncertainty of the location of a particle and the uncertainty of its momentum. Heisenberg saw that the product of these two uncertainties must be bigger than a constant that turns out to be (wait for it) Planck's constant, h, divided by 4 times π. Why Planck's constant is divided by 4π has to do with technical aspects of waves that need not concern us here. The fact that h turns up again and again when describing the atomic regime indicates that Planck's initial guess about the graininess of energy was on target, and by introducing h he discovered a new fundamental constant of nature, as important for understanding how the universe is put together as the value of the speed of light or the charge of an electron.

Right off the hat let me emphasize that the uncertainty principle does *not* restrict the precision with which I can measure the position of a particle, nor that of its momentum. Neither does it state that one can never measure the position and momentum of a particle simultaneously, but it does get to the heart of what such a measurement entails.

Suppose that I can experimentally determine the position of an electron and at the same time the momentum of this electron (I'll describe how in a moment). I may find that the electron is at a location x = 2.34528976543901765438 cm as reckoned from a given point, and the momentum has a value of p = 14.254489765539989021 kg-meter/sec. There are certainly a large number of decimal places in these measurements, but in order to convince myself that this is indeed the electron's position and momentum, I repeat the experiment, under exactly the same conditions. If those are in fact the true values of position and momentum, I should be able to reproduce *all* of those decimal places. Performing the experiment a second time, I now find that the electron's position is x = 2.34528911209876937 cm and the momentum is p = 14.25441008764958320 kg-meter/ sec. On the one hand, the first few digits in both the x and p measurements agree exactly with what was found before, but on the other hand after a few decimal places the overlap with the first observation disappears.

What I find, after repeating the experiment many more times, is that the *average* position of the electron is 2.32428 cm and the *average* momentum is 14.254 kg-meter/sec, as illustrated in Figure 16. That is, the only measurements that one really can trust are of the average position and momentum. I can make a plot of the num-

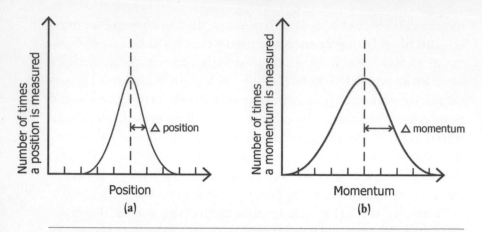

Figure 16: Plot of the histogram of measured positions (a) and momenta (b). The vertical dashed lines represent the average values of position and momentum, and the small arrows indicate the standard deviation for each measurement.

ber of times a particular value of the position is observed against the value of the position. Such a plot would look, after a large number of trials, like the familiar bell-shaped curve, known and feared by students throughout recorded history. The peak in the curve would represent the value of the position that was observed most frequently and would also indicate the average location of the electron. At the highest point, the curve is very narrow. Halfway down the curve, there is a width that is referred to as the "standard deviation." The standard deviation is an indication of how much we can trust the average value resulting from this bell-shaped curve. The bigger the standard deviation, the greater the possible uncertainty in the average value. It is these standard deviations, also referred to in mathematics as uncertainties, that are constrained by the Heisenberg uncertainty principle.

As an illustration of standard deviations, consider a class of 100 students taking a final exam. If all 100 students receive exactly the same score of 50 points out of 100, then the average grade will be 50 and the standard deviation will be 0. That is, if you are told that the average grade is 50 with a standard deviation of 0, then there is no ambiguity or uncertainty in what any given student scored on the exam. Now, if 90 students receive a score of 50 points, while 5 students score a 55 and the remaining 5 receive a grade of 45, the average grade would remain 50, but now there is a small width to the distribution of grades. While the average value is still 50, some students have a score that differs from the average value. There is now some small but non-zero question as to whether a student chosen at random will have a grade equal to the average score of 50 or not. In an extreme case, all 100 students in the class could receive different grades, from as low as 1 or 2, through 48, 49, 50, and 51, up to 98, 99, and 100. The average grade would *still* be 50,* but now the distribution would range from 1 through 100. Only 1 student out of 100 would have an actual grade on the final exam that is exactly equal to the average score of the class, and the large size of the standard deviation would indicate that the average grade was not a particularly meaningful or insightful indicator of any given student's performance. When dealing with large numbers, the two questions one must ask are: What is the average? and What is the standard deviation?

Returning to my simultaneous measurements of an electron's position and momentum, I found that after repeated trials there was a bell-shaped curve for the position, which gave its average location, and another bell-shaped curve for the momentum. It is the standard deviations of these two bell-shaped curves that the Heisenberg uncertainty principle addresses. Heisenberg's theory tells us that the standard deviation of the electron's position is connected to the standard deviation of the electron's momentum, so that changes in one affect the other. The widths of the two distributions in Figure 16 are not independent, and efforts to narrow one will necessarily broaden the other. Heisenberg calculated that the product of the two standard deviations cannot be smaller than $h/4\pi$. Any experiment that manages to decrease the standard deviation of the momentum (for example) will necessarily broaden the standard deviation of the position.

Why would the standard deviation of an electron's position be related to the standard deviation of its momentum? Because of the de Broglie wave associated with the motion of the electron. The wavelength of this matter-wave is in a sense a measure of how precisely we can say where the electron is, and this wave-

* Technically the average is 50.5 in this case—but you get the idea.

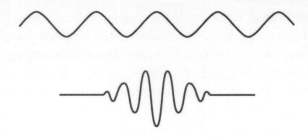

Figure 17: *Sketch of two possible de Broglie matter waves for an electron. In the top curve, the electron is associated with a single wave. As one needs only one wavelength to describe the wave, the momentum is perfectly known, but at the cost of an infinite uncertainty in the location of the electron. In the bottom curve many different waves, each with different wavelengths, have been added to yield a "wavepacket." The uncertainty in the spatial location of the electron is reduced, but there is a corresponding increase in uncertainty in the electron's momentum.*

length is connected to the electron's momentum by the relation (momentum) × (wavelength) = h.

Assume that the electron's momentum is precisely known, with absolutely no ambiguity. Then the average momentum *is* the momentum, just as in our classroom example when all 100 students received the same grade on the final of 50. There is one value for the wavelength of the matter-wave for this perfectly known momentum, as shown in the top cartoon in Figure 17. A pure wave by definition extends forever, from one end of the universe to the other. Where would we say the location of the electron would be for such a perfect wave? Its average value might be well defined, but its standard deviations would be infinitely large.

The Heisenberg connection between the standard deviations of the position and momentum results from the fact that the location of the electron is connected to the wavelength of the matter-wave, which in turn is related to the object's momentum. In order to shrink the standard deviation of the electron's position, the electron's matter-wave should be zero except for a small region near the average position. But in order to construct a wave packet such as this, illustrated by the lower sketch in Figure 17, one needs to add many waves together, all of slightly different wavelengths, so that they would destructively cancel out beyond a narrow region around the average position. Since the wavelength is connected to the momentum through (momentum) × (wavelength) = h, adding many different wavelengths together is the same as saying that there is a broad range of momentum values for the localized electron. The tighter the limits with which we wish to specify the electron's location, that is, the smaller the position's standard deviation, the more different wavelengths we have to combine and the greater the corresponding standard deviation of the momentum. The two standard deviations are thus joined together, through the matter-wave relation (momentum) × (wavelength) = h.

Of course, we can measure an electron's position to any arbitrary precision we wish—as long as we do not care about its momentum, and vice versa. Whether the electron has a well-defined momentum (with no standard deviation) or a well-defined position (also with no standard deviation) depends on what measurement I perform. There can be no answer until the question is posed, and how I ask the question determines what answer I'll obtain.

The bundle of waves described above, and shown in the lower sketch in Figure 17, is technically referred to as a "wave packet." It makes no sense to ask where the electron is located, on a scale smaller than the extent of the wave packet. How do I measure the momentum of an electron? One way would be to record its position at two different times. Knowing how far it moved, and how long it took to travel this distance, I can determine its velocity, and multiplying by the mass (assuming I can avoid relativistic corrections) yields the momentum. How can I tell where the electron is at the two different times? It's easy to tell the velocity of an automobile— you just look at where it is at two different times. How do I look at an electron? The same way I look at a car—by shining light on it and having the light be reflected to my eye (or some other detector). Cars are much larger than the wavelength of visible light, but to observe the electron I need light with a wavelength smaller than the extent of the electron's wave packet. The connection between wavelength and frequency for light is given by the simple relationship (frequency) × (wavelength) = speed of light. Since the speed of light is a constant, the smaller the wavelength, the larger the frequency of the light, and from energy = h × (frequency), the larger the energy of the light photon. Thus, to measure the momentum of an electron with a very small wave packet (small uncertainty in position), I must strike it with a very high-energy photon. When the photon reflects from the electron, the electron's recoil changes its momentum. When a careful mathematical analysis is performed, one finds that you cannot do better than the Heisenberg uncertainty principle. You may think there might be a clever scheme to get around this limitation of the wavelength of light to determine the electron's position and momentum—but many have tried and all have failed.

We might have been spared countless inane pronouncements that "quantum mechanics has proven that everything is uncertain" if Heisenberg had simply named his principle something a little *less* catchy, such as "the principle of complementary standard deviations."

Armed now with an appreciation for the physical content of the famous uncertainty principle, we now consider the following classic example of nerd humor:

> Werner Heisenberg is pulled over for speeding by a highway patrolman. The police officer walks over to Werner's car, leans over, and asks Heisenberg, "Do you know how fast you were going?"
> Heisenberg replies, "No, but I know where I am!"

*   *   *

An understanding of average values and standard deviations of bell-shaped curves is also relevant to issues of climate change and explains why scientists are concerned about an increase in the average global temperature of a few degrees, which admittedly does not sound that menacing.

Consider a histogram plot of the temperature of North America, where the horizontal axis lists the average daily temperature, and the vertical axis charts the number of days per year for which that particular average temperature is recorded. We expect to find, and we do, a roughly bell-shaped curve, just as we have been discussing for grades in a large class or measurements of an electron's position or momentum. There will be a large peak in the temperature histogram around the average daily temperature over the course of a year, and small tails at lower and higher temperatures.

What is the effect of the yearly average temperature being raised by a few degrees? The answer is easiest to see if we briefly return to the classroom analogy. Suppose we add 5 extra-credit points to every student's final exam paper. Thus, if the average had been a grade of 50 before, now it will be 55, and the lowest score will shift from 0 to 5, while the highest possible grade will move from 100 to 105. A shift in the average grade from 50 to 55 does not seem like much, and for most students there will be no significant effect. I am just adding a uniform 5 points to everyone's grade, so the shape of the curve does not change. In addition, I do not adjust the cutoff line for what grade merits an A, what score deserves a B, and so on. In this case I would find that by shifting the grades upward by only 5 points, the number of students that are at or above the A-cutoff threshold has increased dramatically. A minor shift in the average, which does not have a large influence on the grade of the majority of the students, has a big impact on those students that were near but just below the threshold for a letter grade of A.

A small shift in the average annual global temperature is akin to giving everyone in the class 5 extra-credit points. There will thus be more days with higher-than-normal average temperatures (corresponding to those students whose exam performance warranted an A), and those days are what drive extreme weather situations. A shift of the average upward by a few degrees is not a big deal on an average day and is even welcome in the winter in states such as Minnesota, where we would not have such extreme cold snaps. But other parts of the nation in the summer would see a greater number of days where it is hotter than normal. It takes a lot of energy to warm up a large mass of water such as the Gulf of Mexico, and it also takes a long time to cool it back down. The energy of these hotter-than-normal days can be viewed as "stored" in the ocean, and warmer water temperatures can provide energy for hurricanes, tropical storms, and other extreme events. Moreover, the more days with higher temperatures, the more ice will melt in northern regions. Fresh snow reflects 80 to 90 percent of all sunlight shining on it, while liquid water absorbs (and stores) 70

percent of the sunlight. There is thus a positive feedback mechanism by which higher temperatures lead to additional warming.
Just as in the case of the Heisenberg uncertainty principle, it's not
the averages that matter so much as the width of the standard deviations. Those long tails will get us in the end.*

---

* I apologize for that joke.

CHAPTER EIGHT

Why So Blue, Dr. Manhattan?

In Chapter 5 I stated that at least some of the amazing superpowers displayed by Dr. Manhattan in the graphic novel and motion picture *Watchmen* are a consequence of his having control over his quantum mechanical wave function. Now that we know a little bit more about wave functions, let's see how that might work.

While it is certainly true that all objects, from electrons, atoms, and molecules to baseballs and research scientists, have a quantum mechanical wave function, one can safely ignore the existence of a matter-wave for anything larger than an atom. This is because the larger the mass, the larger the momentum—and the bigger the momentum, the smaller the spatial extent of the wave function. Anything bigger than an atom or a small molecule has such a large mass that its corresponding de Broglie wavelength is too small to ever be detected. So, right off the bat we must grant Dr. Manhattan a miracle exception from the laws of nature such that he can control his wave function's spatial extent independently of his momentum. While the de Broglie wavelength for an adult male is typically less than a trillionth trillionth of the width of an atom, Dr. Manhattan must be able to vary his wave function so that it extends a great distance from his body—even as far as the distance between the Earth and Mars!

The quantum mechanical wave function contains all the information about an object. If we want to know the object's average position, its average speed, its energy, its angular momentum for rotation about a given axis, and how these quantities will change with time, we perform various mathematical operations on the wave function, which yield calculated values for any measured characteristic of the object.

The "wave function" is so named because it is a mathematical function that has the properties of an actual wave. To review: In mathematics a "function" describes any situation where providing one input value leads to the calculation of a related number. The simple equation relating distance to time spent driving at a constant speed—that is, distance = speed × time—is a mathematical function. If your speed is 60 miles per hour, and if you tell me the time you spent driving—1/2 hour, 1 hour, 3 hours—then this simple function enables me to calculate the distance you have covered (30 miles, 60 miles, or 180 miles, respectively, in this example). Most mathematical functions are more complicated than this, and sometimes they get as involved as the Schrödinger wave function, but they all relate some input parameter or parameters to an output value. For the quantum mechanical wave function of an electron in an atom, if you tell me its location in three-dimensional space relative to the nucleus, then the solution to the Schrödinger equation returns the amplitude of the electron's wave function at that point in space and time.

What does it mean to say that the wave function has the properties of a wave, such as a vibrating string or the series of concentric circles created on the surface of a pond when a rock is tossed into the water? Waves are distinguished by having amplitudes that vary periodically in space and time. Consider the ripples created when a rock is tossed into a pond. At some points of the wave there are crests, where the height of the water wave is large and positive (that is, the surface of the water is higher than normal); at some points there are troughs, where the height of the water is lower than normal; and in other regions the amplitude of the wave is zero—the height of the water's surface is the same as it would be without the rock's disturbance.

The amplitude of the peaks and valleys typically becomes smaller with distance from the source of the waves. This is why on the California shore we don't notice if a rock is dropped into the center of the Pacific Ocean. Certain large disturbances can create tsunamis that maintain large amplitudes even when traveling great distances. Dr. Manhattan, presumably, is able to change the amplitude of his quantum mechanical wave function so that it can have an appreciable amplitude at some large distance away from him. This would be how he teleports, though in quantum mechanics we would say that he is "tunneling."

Schrödinger's equation enables us to calculate the wave function of an object as a function of the forces acting on it. If there are no net forces, the electron, for example, can have uniform straight-line motion, with a well-defined de Broglie wavelength determined by its momentum. If this electron strikes a barrier and it lacks sufficient energy to go over the obstacle, then the electron will be reflected, bouncing off the barrier and returning from where it came.

We are familiar with such wave phenomena whenever we use a mirror. Light waves move in straight lines, passing through the glass covering of the mirror, until they reach the silvered backing. Unable to penetrate the metal, the waves are reflected back in another straight-line trajectory, along a path that makes the same angle with a line perpendicular to the mirror's surface as the incoming beam.

In fact, one does not need the metal backing to see this reflection effect. We all know that a single pane of glass can act like a mirror, when we look out the window from a well-lit room at night. In this case just the difference in optical media, glass and air, can cause light reflection, particularly when we look at the window at an angle. The reflection is more noticeable if the direction we are looking, relative to the glass surface, is larger than a particular angle that depends on the optical properties of glass and air. When we place our face against the glass, this reflection effect goes away, for then most of the light rays that we see from outside travel perpendicular to the surface. Light travels slower in glass than in air (more on this in a moment), and this difference in light velocities (characterized by the material's index of refraction, for technical reasons) accounts for the reflection effect. This can occur during daytime as well but is less noticeable when more light comes into the room from the outside than goes out from the interior.

Suppose that we are looking at the window at night from the interior of a strongly lit room. The glass reflects our image as if it

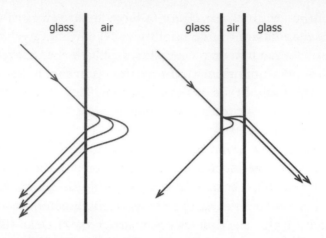

glass   air   glass   air   glass

Figure 18: *Cartoon sketch illustrating a light wave, which is normally reflected at a glass/air interface and may have a small amplitude leaking into the air. If another piece of glass is placed near the first (the separation should be no more than a few wavelengths of the light), then the wave may be able to propagate into the second material. A similar phenomenon occurs with matter waves during quantum mechanical tunneling.*

were a conventional mirror. Now imagine a second sheet of glass placed behind the first, as in a double-paned window, only the separation between the two sheets isn't a quarter of an inch, but more like a millionth of a centimeter. In this case, even though the light would have been completely reflected without the second sheet of glass, the presence of the second pane enables some of the light to pass through both sheets of glass, even though they are not touching each other. This phenomenon is a hallmark of the wavelike properties of light (so for the sake of argument we will ignore for the moment that light is actually comprised of discrete photons). It turns out that the light wave is not completely reflected at the first glass-air interface, but instead a small amount of the oscillating electric and magnetic fields leak out into the air. The small leakage is limited to a thin region very near the interface and is normally not important. But when the second sheet of glass is brought very close to the first interface, some of the protruding light waves extend into the second piece of glass. In this case the wave is not reflected but rather propagates into the second glass.

This "leakage effect" is not unique to light; it occurs for any wave—even those associated with matter!

One of the most fantastic aspects of quantum mechanics, and one that provides dramatic confirmation that there is a wavelike aspect to the motion of matter, is that this "leakage effect" is observed for electrons, protons, and neutrons. Here, instead of light and a sheet of glass, consider an electron in a metal or semiconductor. Instead of a glass-air interface, there might be a barrier at the surface of the conductor, either the vacuum of empty space or some other electrical insulator. The electron would normally not have enough energy to leave the conducting material and would be reflected at this surface. If another conductor is placed on the other side of the barrier, and if this barrier is not too thick compared to the electron's de Broglie wavelength, then there is a probability that the matter-wave can extend through the gap. Even though the electron does not have sufficient energy to jump or spark across the gap, as its quantum mechanical wave function leaks through the forbidden barrier into the second region, it can thus be found in the second conductor.

When matter waves exhibit this leakage effect, it is termed "quantum mechanical tunneling" even though the electron does not, obviously, create a "tunnel" through the insulator or the vacuum of empty space. Recall that the square of the wave function tells us the probability per volume of finding the electron at some point in space and time. If the leakage of the wave function through the forbidden region is small, then there is a low probability of finding the electron in the second region on the other side of the barrier. But anything that has a probability larger than zero will happen if one tries enough times. If we send an electron moving toward the barrier of a particular height and width, examination of the wave function may show that there is a very large probability—at least 99.9999 percent—that the electron would be reflected at the interface. This means that there is only one chance in a million that the electron will show up in the second material. But if a million electrons approach the barrier, one may get through, and if a trillion electrons strike the barrier, then a million will pass through via tunneling to the other side. We don't know which electrons will make it into the second conductor until we send them out, but

Figure 19: An example from the graphic novel Watchmen—*Dr. Manhattan's physics remains unchanged when he multiplies his quantum mechanical wavefunction by a constant bigger than unity.*

based on the properties of the barrier, we can confidently predict how many on average will get through. As discussed in Section 5, many personal electronics devices employ the tunneling phenomenon to regulate the current in a circuit—putting this esoteric quantum mechanical effect to prosaic and reliable use.

What an electron can do, Dr. Manhattan can do as well, at least in the pages of a comic book or a motion picture with an extensive special-effects budget! Presumably through his miraculous control of his quantum mechanical wave function, Dr. Manhattan is able to extend his de Broglie wavelength not just a few nanometers, as the electron in a tunneling diode does, but over thirty-six million miles.* With a large enough amplitude at the remote location, the probability of Dr. Manhattan suddenly appearing at the new site becomes very large. He never is actually in the space between his starting point and final destination but is simply able to adjust his probability density to be a maximum where he wants to go—which is certainly a savings in time and money compared to commercial air travel.

Dr. Manhattan is able to change his size at will (as shown in Figure 19) due to the fact that the Schrödinger equation is linear. In

---

* At Mars's closest point to our planet.

mathematics an equation is called "linear" if it depends only on the key variable (in the Schrödinger equation that would the wave function Ψ) and not on that variable squared or cubed, or the square root, and so on.* A very simple linear equation is $\Psi = \Psi$, which is certainly a true statement. In fact, this equation is so simple that it is always true for *any* value of Ψ. So if $\Psi = 1$, then this equation tells us that $1 = 1$ (which we already knew). In this case, if Ψ is ten times larger, then this simple equation tells us that $10 = 10$, which is also a true statement. Given that the Schrödinger equation is linear, there is no change in the physics of the situation if we multiply Ψ by a constant, either a larger or smaller one. By multiplying the wave function by a constant (the "normalization" described in Chapter 6), we ensure that Ψ^2 acts as a probability density and varies from 0 percent to 100 percent. In addition, the fact that the Schrodinger equation is linear means that if there are two possible solutions to the equation, such as Ψ_A and Ψ_B, then their sum $\Psi_A + \Psi_B$ will also be a solution (this will be very important in Section 4). Presumably Dr. Manhattan is able to shrink himself down as well, multiplying his wave function by a value less than 1, though we never see him utilize this capability in the comics or the motion picture adaptation.

Jon Osterman, as shown in Figures 11 and 19, gained a bright blue pallor when he reassembled himself following the unfortunate "incident" in the intrinsic field chamber. As wave functions have no color, there are at least three possible explanations for his being blue: (1) always knowing what will happen in the future has taken all the joy out of life; (2) he's depressed because he realizes that "nothing ever ends"; or (3) he's emitting Cerenkov radiation.

Dave Gibbons, the artist of *Watchmen*, once stated in a radio interview that he elected to make Dr. Manhattan blue as a visual signifier in order to constantly remind the readers of Jon's transformation. If Dr. Manhattan were red he would look like he was on fire, green was too close to the Hulk, and other colors would look too similar to actual skin tones on the printed comic page. Be that as it may, just because the color choice was one of casual necessity

---

* A nonlinear equation would be something like $\Psi^2 = 2\Psi$, which is a true statement for $\Psi = 2$, but not for a Ψ value ten times bigger, that is, $\Psi = 20$.

Figure 20: *Image of a pencil (belonging to a certain fictional physicist) that appears broken at the air/water interface due to the different speeds of light in the two media.*

does not mean that we can't obsessively discuss the underlying physics in great detail! For it turns out that given Dr. Manhattan's origin, if he were to glow in any color of the optical spectrum, it would indeed be blue.

When certain elements undergo radioactive decay, they may emit high-speed electrons as a by-product of their nuclear reaction (we'll discuss the mechanism by which this occurs in the next section). When those electrons (also referred to as "beta rays") travel faster than the speed of light in a material medium, they emit electromagnetic radiation in the blue-ultraviolet portion of the spectrum, which is known as Cerenkov radiation.

This last sentence is no doubt puzzling, for a central principle of Einstein's Special Theory of Relativity is that nothing can travel faster than the speed of light. But this is in fact not strictly correct. The more accurate way to state this principle is that nothing can travel faster than the speed of light—*in the vacuum of empty space*! Light speed in a vacuum is three hundred million meters per second and is indeed the fastest velocity in the universe. However, light travels much slower than this when moving through denser media, such as water or glass.

Anyone who has noted that a straw or pencil in a glass of water appears to be "broken" at the water-air interface, as shown in Figure 20, has observed an optical effect that results from light moving slower in water than in air. In order to be seen, light must be reflected from the straw and detected by our eyes. The change in the speed of light at the water-air surface causes straight-line light rays to bend, in a phenomenon termed "refraction." The light that bounces off the portion of the straw protruding from the water of course does not bend and travels in a straight line. When we observe the light from the straw in the air and the light that bent upon leaving the water, we interpret the image as a straw with a sharp discontinuity at the water's surface.

Why does light travel slower in water and other media? It is because the electromagnetic waves interact with the electrons surrounding each atom in the material. When running through a swimming pool, you will move slower if you hold your arms out away from your body and increase the drag from the water. Light experiences an "electromagnetic drag" from the electrons that can slow its motion down markedly. The speed of light in water or glass is only 75 percent of what it is in a vacuum, which is still pretty fast. But high-speed electrons can move through these media with fewer interactions, and thus it is possible for an electron to travel in water faster than light can. When this happens, the electron (which does interact with the electrons surrounding the atoms in the material, only not as strongly as light does) generates an "electromagnetic sonic boom," emitting light in the blue-ultraviolet region of the electromagnetic spectrum. This blue-light shock front is termed Cerenkov radiation, after Pavel Cerenkov, who discovered and explained this phenomenon in 1934 (for which he was awarded the Nobel Prize in Physics in 1958).

Air is much less dense than water or glass, and light slows down only slightly when moving through the atmosphere compared to its largest speed in a vacuum. Nevertheless, for the purposes of explaining the science underlying a fictional character in a comic book, let's stipulate that it is possible to generate Cerenkov radiation from high-speed electrons jetting through the air. Let's also suppose that when Dr. Manhattan reassembled himself following the removal of his intrinsic field, he did so in such a way that he is continually leaking high-speed electrons, giving him a healthy blue glow. There are always many electrons from the Earth that he can draw upon in order to maintain his charge neutrality. If he wanted to darken his hue (as he does at one point for the benefit of television cameras), he could simply change the speed at which the electrons escape.

Nuclear reactor piles at the bottom of deep pools of water* give off a blue glow, and this Cerenkov light indicates that the pile is active and emitting beta rays. In *Watchmen* (spoiler alert!) a character frames Dr. Manhattan, so that he is accused of giving his close friends and an ex-girlfriend cancer. One way to inflict Osterman's associates that would plausibly suggest him as the source of the disease is to surreptitiously expose these people to nuclear isotopes, such as strontium-90, that are known to be carcinogenic and are deadly precisely because of their beta radiation emissions.

Another striking characteristic of *Watchmen*'s Dr. Manhattan is his ability to experience the past, present, and future simultaneously. It is specified in the graphic novel that the post-intrinsic-field-removal Jon Osterman is able to see only his own future and thus would not know of events to come unless he either directly experiences or participates in them or is told about them. Again, if Dr. Manhattan did indeed have control over his macroscopic quantum mechanical wave function, then as the wave function contains *all* the information about the object's probability density in space *and* time, this characteristic is plausible.

The fact that there is no other source of information about the future evolution of an object than what is contained in its wave function is significant. If all we have is the wave function, and the wave function can tell us only the probability per unit volume of finding the object in space and time, then, even in a perfect, idealized situation, we must resign ourselves to knowing only the odds as to the object's location. When we deal with probabilities and statistics in other nonquantum situations in physics, it is simply to make our lives easier. We know that Newton's laws of motion provide a nearly complete description of the interactions of the air molecules in the room in which you are reading this right now. However, to apply these equations to the air would involve

---

* The heat emitted by the reactor causes the water to boil, and the resulting steam turns turbines in the dynamo that generates electricity.

solving Newton's laws for all trillion trillion molecules simultane-ously. In this and similar situations, it is much more reasonable to describe the average pressure, for example, or introduce the con-cept of "temperature" (which represents the average kinetic energy per molecule) rather than deal with each molecule separately in turn. In contrast, in the quantum world, the emphasis on probabil-ity density is a matter of necessity, not convenience. Even with infinitely fast and infinitely precise observations, we can never know exactly where the object is, but only its average location.

This inability to do better than knowing the odds is a conse-quence of the wavelike nature of matter. Recall the discussion of the Heisenberg uncertainty principle from the preceding chapter. The wavelength of the matter-wave associated with the electron, for example, is directly connected to its momentum. A pure, single wave has only one wavelength, and thus we know exactly what its momentum is, but at the expense of having any information about where the electron is. The more we localize the electron, say, by ensuring that it will be found within the one-third of a nanometer that is the typical spatial extent of an atom, the less defined its momentum becomes. If we had perfect knowledge of its position (which is what physicists desired in order to put the "probability density" aspect to rest), then this would come at the cost of perfect ignorance about its momentum. It could in principle have any mo-mentum between zero and infinity, and we would thus have to contend with a probability interpretation of its motion. As we need both positions and momenta to employ a traditional Newton's law description of a system, probabilities are the best we can ever do.

Of course, knowing the probability that Dr. Manhattan may be in a particular state in the future, such as on Mars having a conver-sation with his girlfriend, is not a guarantee that he will indeed work out his relationship problems on the red planet. The only time something is absolutely certain to occur is when the probabil-ity is 100 percent, just as the only time something will never hap-pen is if the probability is zero.

In most circumstances the most probable outcome is indeed the one that is observed. But what about the other probabilities that are not realized? What do these wave function solutions to the Schrödinger equation correspond to? One interpretation was provided by Hugh Everett III. Everett suggested that all these probabil-ities describe actual outcomes on other Earths in an infinite number of parallel universes! If the probability of a certain event occurring is 10 percent, then Everett suggested that on 10 percent of the pos-sible parallel Earths this outcome did indeed occur. The world we live in and experience is one that continually unfolds from this multiverse of possible Earths. For everything we experience, there are alternate Earths where different outcomes are realized.

Everett's ideas were considered too unconventional even by the standards of quantum theory, and his proposal, described in his physics dissertation at Princeton in 1957, earned him his Ph.D. but was otherwise completely ignored by the scientific community. Disappointed, Everett eventually turned away from pure scientific research and worked for the military, calculating fallout yields of various nuclear weapons for the Department of Defense. He passed away in 1978, but not before his ideas received some measure of recognition by a small group of theoretical physicists, notably Bryce DeWitt, who actually coined the term "many-worlds inter-pretation of quantum mechanics" to describe Everett's thesis. Now-adays the number of physicists who subscribe to the many-worlds picture, while still small, is growing, as those who are struggling to reconcile quantum mechanics and Einstein's General Theory of Relativity find application for the many-worlds model.

Parallel universes and alternate Earths are, of course, a com-mon feature in science fiction stories, both prior to Everett's dissertation and since. Sometimes these alternate worlds are pro-foundly different from ours, as in *Flatland*, Edwin Abbott's tale of a two-dimensional world published in 1884, or the 1931 short story "The Fifth Dimensional Catapult," by Murray Leinster. In 1896 H. G. Wells told "Plattner's Story," wherein Gottfried Plattner, in an accident involving a mysterious green powder in a chemistry lab at a boys' boarding school, is hurled to a parallel world that orbits a green sun and is inhabited by strange alien creatures with human heads and tadpole-like bodies. It is difficult to imagine the branching of possible wave functions that could have led to such an outcome. In Wells's short story "The Remarkable Case of Da-vidson's Eyes," Sidney Davidson, through another laboratory ac-cident, gains the ability to see another world, where a ship docks on a South Sea island and stocks up on penguin eggs, despite the fact that all the information from his other senses is consistent with his being in a laboratory in London. Gradually Sidney's normal vision returns, and in time he discovers that the ship that he had seen in this alternate Earth was a real sea vessel that was in fact gathering penguin eggs on Antipodes Island at the time of Davidson's strange visions. While a definitive explanation is not presented, it is speculated that when Davidson stooped between the poles of a powerful electromagnet in the lab, his retina gained the ability to see through "a kink in space"—though whether of this world or a parallel one remains open to interpretation.

A few years after Everett published his novel solution to the "measurement problem" in quantum mechanics, the DC super-speedster the Flash of the 1960s vibrated to a parallel Earth and had an adventure with the Flash of the 1940s (same power, different costume and alter ego). In the television program *Star Trek* broadcast in 1967, a transporter malfunction during an ion storm leads Captain Kirk, Dr. McCoy, Engineer Scott, and Lieutenant Uhura to an alternate universe starship *Enterprise*, populated by evil twins of the rest of the crew (distinguished by goatees, naturally). In this mirror universe, the crew of the *Enterprise* are violent and ruthless, but one feature that remains constant in either universe is Captain Kirk's roving eye for the ladies.

In comic books, characters often travel to alternate Earths in parallel universes, and the implication in the stories is that the world of the comic book reader, the one lacking in actual superheroes, is the "real universe." However, a photo that I came across in the archives of the American Institute of Physics suggests that the situation may be more complicated than we might think. The photo, shown in Figure 21, documents a visit in 1954 to the Princeton University physics department by Niels Bohr (one of the founders of quantum mechanics we encountered in Section 1) as he meets with several physics graduate students. The student on the immediate right of Bohr is Hugh Everett III. The student on the far left appears to be none other than Jon Osterman! Recall that Osterman received his Ph.D. in physics from Princeton in 1957 and so would have indeed been included in the select group of students honored with an audience with one of the grand old men of phys-

Figure 21: *Niels Bohr (center) visiting with some physics graduate students at Princeton University in 1954. Second from the right, to Bohr's immediate left, is Hugh Everett III, who would posit the existence of an infinite number of Earths in parallel universes in order to resolve the "measurement problem" in quantum mechanics. At the far left is Charles Misner, a graduate student with a resemblance to Jon Osterman (inset), who would become Dr. Manhattan in* Watchmen.

ics. As mind-bending as the concepts introduced by quantum mechanics into modern thought have been, the suggestion that comic book characters live among us may be a step too far!*

---

* And indeed, this student is Charles Misner, who has made many important contributions to the study of cosmology and gravitation over his distinguished career.

SECTION 3

TALES OF THE ATOMIC KNIGHTS

CHAPTER NINE

Our Friend, the Atom

In the 1949 Warner Bros. musical motion picture *My Dream Is Yours*, a young Doris Day auditions for a spot as a featured singer on a popular radio show. Her manager, played by Jack Carson, advises her to curry favor with the sponsor by crooning a tender love ballad. She instead decides to belt out a bouncy, up-tempo ode to a "new invention . . . [no] larger than an adding machine . . . [that] few have ever seen." As the song continues, joined by the refrain of "tic, tic, tic," it becomes clear that Doris Day is singing the praises of—and comparing her quickly beating heart to—a Geiger counter!

Five years later, a ticking Geiger counter in another film would lead uranium prospector Mickey Rooney onto an atomic bomb testing site. Not realizing that a nuclear weapon detonation was imminent, he innocently took refuge in a test house populated with mannequins and helped himself to a peanut butter sandwich as the countdown progressed. Rooney survived the nuclear explosion without having to take refuge inside the model refrigerator. The resulting exposure to radioactivity would transform Mickey Rooney into *The Atomic Kid*, and he would go on to employ his newfound ability to glow in the dark and issue explosive sneezes to help the FBI break up a communist spy ring.

A few years after Mickey Rooney's misadventures on a nuclear weapon testing site, a darker though equally inaccurate depiction of the effects of radiation exposure would be presented in *The Beast of Yucca Flats* (1961). In this cautionary tale, former Swedish wrestler Tor Johnson (of *Plan 9 from Outer Space* fame), also accidentally wanders into an atomic bomb test run. Johnson plays defecting Russian scientist Joseph Javorsky, who, while fleeing KGB assassins, winds up on the famous desert Yucca Flat testing range right before an atomic bomb detonation. The resulting radiation transforms Johnson in a hulking, mindless homicidal monster (though he looks pretty much the same as before the explosion).

Certainly the true effects of radiation exposure were publicly known at least by August 1946, with the publication in the *New Yorker* of John Hersey's "Hiroshima." But in the years immediately following the conclusion of World War II, popular forms of entertainment maintained, for the most part, an optimistic view of the benefits to come in an atomic-powered world of tomorrow. The 1957 television program *Disneyland* featured Dr. Heinz Haber, a German rocketry expert, in *Our Friend, the Atom*, which likened atomic power to a genie in a bottle that could grant us three wishes for a brighter future. The first wish would be for power, from the generation of electricity to atomic-powered airplanes. The second wish was for food and health and involved using radiation to sterilize foodstuffs and in the treatment of diseases. The third wish was for wisdom, to use nuclear energy wisely and peacefully.

In 1952 *Collier's Magazine* commissioned a series of articles by science writers from Wernher von Braun and Heinz Haber to Willy Ley to envision the future of space travel. With illustrations by Chesley Bonestell, who did the background artwork for *Destination: Moon*, and Rolf Klep, these articles were published as three issues of the magazine and later compiled into book form under the title *Across the Space Frontier*. Here again, the "genie" of atomic energy would provide the power to run space stations and enable manned missions to Mars. Before the grim realities of mutated Swedish former wrestlers set in, there was a real sense of optimism—that the taming of the atom and our understanding of nuclear physics would make the promised utopias of science fiction a reality.

What went wrong? While we fortunately avoided glow-in-the-dark Mickey Rooneys, we never got the atomic planes either. Well, the atomic planes were a bad idea from the start. Haber's Dell paperback companion to the Disneyland television program argued, "In aviation, the weight of the fuel has always been a discouraging limitation." (Now it's the *cost* of the jet fuel. But back in 1956, no one envisioned the end of cheap oil.) While a smaller nuclear reactor can replace a large quantity of fuel, the shielding necessary to prevent killing or sterilizing the passengers and crew would more than compensate for the missing fuel weight. Haber suggested using water as shielding, but the now heavier plane would require a runway miles long—all of which hardly seems worth the trouble simply to be able to avoid refueling on long flights.

Similarly stillborn were plans for atomic automobiles. In 1957 Ford proposed a car called the Nucleon,* in which the internal combustion engine would be replaced by a small nuclear reactor located in the back trunk. The heat from a nuclear fission reaction would boil water, and the steam would turn turbines, providing torque for the wheels and electrical power, as in a nuclear electrical power plant. The hazard to the driver from exposure to nuclear radiation, and to other motorists from a traffic accident, was to be offset by the improved mileage—it was anticipated that the Nucleon could travel five thousand miles before the atomic core needed replacement. Though never built, the three-eighths-scale model unveiled by Ford is notable for a mini–cooling tower behind the passenger section for the nuclear reactor and tail fins nearly as tall as the car itself.

Certainly the benefits of atomic-powered travel outweigh the costs when considering underwater transportation. The first U.S. Navy nuclear-powered submarine, the *Nautilus,* was launched in 1954, and since then a considerable fraction of the global fleet of submarines is powered by small nuclear reactors. The *Nautilus* in Jules Verne's *Twenty Thousand Leagues Under the Sea* was powered by electricity drawn from the ocean, via a mechanism not clearly described ("Professor," said Captain Nemo, "my electricity is not everybody's and that is all I wish to say about it. . . . "), and consequently was also able to travel great distances without refueling (twenty thousand leagues refers to the distance the *Nautilus*

---

* A "nucleon" is the term physicists use for a particle inside a nucleus, either a proton or a neutron.

travels, not its depth beneath the water's surface, and is equivalent to sixty thousand miles). As the sole market for submarines is the military,* profitability constraints do not apply.

It is true that nuclear power is extremely efficient compared to other methods of generating heat, at least when compared to the equivalent mass of fossil fuel needed to produce the same energy. The devil is in the details—particularly in the waste products. While there is danger in the waste exhaust of fossil fuels, there the hazard is long-term, while radioactivity is of immediate concern to all it strikes. To see why we must be concerned when a nucleus decays, we first need to understand why any nucleus sticks together in the first place.

When Ernest Rutherford's lab conducted experiments involving high-speed alpha particles (consisting of two protons and two neutrons, essentially a helium nucleus) scattering from thin metal foils, they observed that occasionally, say one time in ten thousand, the alpha particles were reflected backward from the metal foil. These data led them to conclude that the atom was mostly empty space (which we now understand to be occupied by the "probability clouds" for the electrons) and a small inner core in which the positive charges reside. The positive charges have to be in the center, for only a concentrated volume of positive charge could generate a repulsive force sufficient to deflect the high-velocity alpha particles (which themselves contain two positive charges) backward from their initial trajectory. This nucleus had to be small, in fact, roughly one ten-thousandth the diameter of the atom itself, in order to account for the fact that only one in ten thousand alpha particles experiences a significant deflection (as a direct hit is necessary to send the alpha reeling backward).†

Knowing that the positive charges in the atom were in the nucleus answered the question of the structure of the atom but raised several more. It was known from chemistry that the number of positive charges in an atom (balanced by an equal number of negatively charged electrons) determined its chemical nature. Hydro-

---

* The sponsor for the Disneyland program *Our Friend, the Atom* was General Dynamics, manufacturer of, among other things, nuclear submarines.

† Assuming that the alpha particle is electrically repelled before striking the nucleus.

gen has one proton in its nucleus, helium has two, carbon has six protons, while gold has seventy-nine. The electron's mass is nearly two thousand times smaller than a proton's, so nearly all of the mass of the atom derives from its nucleus. But the weight of an atom does not correspond to the number of net positive charges it has. Hydrogen has a mass equivalent to a single proton, but helium's mass is equal to that of four protons, carbon's is twelve, and gold's mass would suggest that it has 197 protons in its nucleus.

How can helium have a nucleus with only two positive charges, but a mass four times larger than that of hydrogen? For a while, physicists thought that the nucleus contained both protons and electrons. That is, a helium nucleus would consist of four protons and two electrons. That way, it would have a mass four times larger than hydrogen's single proton, as observed, but a net charge of $+4-2 = +2$, which also agreed with the experiments. As the electron has a much smaller mass than the proton, measurements at the time were not precise enough to rule this possibility out.

Experiments on the nuclear magnetic field (remember that protons have small magnetic fields, as discussed in Chapter 4) and how it influenced the manner by which the electrons in the atom absorbed light (more on this when we discuss magnetic resonance imaging) led scientists to conclude that a helium nucleus, for example, could not have four protons and two electrons. Instead there must be two protons in a helium nucleus, and two other particles that weigh as much as a proton but have no electrical charge. In 1932, James Chadwick bombarded beryllium with alpha particles and detected a new part of the atom: the neutron. Thus one mystery about the nucleus was solved—the atom consisted of electrons orbiting a nucleus that contained protons and neutrons.

But this left another, more challenging mystery. As it is well known that like positive charges repel one another (this was, after all, the basis by which Rutherford had discovered the nucleus—by observing it repel positively charged alpha particles), then why do the positively charged protons in the nucleus not fly away from one another? The answer is—they do! Protons "feel" electrical forces inside the nucleus just the same as outside the nucleus. The fact that they stay inside the small nuclear volume implies that they feel a second, stronger force that prevents them from leaving the nucleus. A clue about this force is found by considering the heavier siblings of each element, termed "isotopes." Two atoms are isotopes if their nuclei have the same number of protons (thus making them identical chemically) but differing numbers of neutrons (thus giving them different masses). There are versions of hydrogen that have one proton and zero, one, or two neutrons,* but there are no isotopes of helium or any other element that have two or more protons and no neutrons. This indicates that the neutrons in the nucleus play a crucial role in providing the "strong force" that holds the nucleus together (the same strong force we encountered in Chapter 5).

How much stronger is this force than electromagnetism? If this additional force were ten times greater than the electrical repulsion, then it would be hard to make heavy elements such as silicon, with fourteen protons, or titanium, with twenty-two protons. If the force were a thousand times stronger, then we might expect to see elements with several hundred protons in the nucleus, and we do not. The fact that the heaviest natural element found on Earth is uranium, with ninety-two protons, indicates that this strong attractive force holding the nucleus together is roughly one hundred times greater than the electrical repulsion between the protons.

But even uranium is not stable, and if you wait long enough, all of your uranium will undergo transmutations to smaller elements by a process known as radioactive decay. Lead, with fifty-six protons and 126 neutrons, is the largest element that does not decay and is therefore stable. You can construct heavier nuclei, but when the "tower of blocks" of protons and neutrons becomes too tall (for each additional proton means more neutrons have to be present to keep it together), eventually the slightest perturbation will cause the tower to collapse. When it does, it loses energy by emitting radiation in the form of high-energy photons (gamma rays) or high-speed subatomic particles, such as electrons, neutrons, or alpha particles.

In fact, some of the larger nuclei are so unstable that all you

---

* To maintain stability in a nucleus requires a critical balance of the number of neutrons and protons—consequently, isotopes such as hydrogen, with one proton and two neutrons, may be unstable and "decay." More on this soon.

have to do is give them a tap, and they fly apart. Uranium, so valuable in the middle of the 1950s that it would tempt Mickey Rooney out into an atomic testing site, is one such element. A dictionary from the end of the nineteenth century described uranium as "a heavy, practically worthless metal." But this was before Otto Hahn and Fritz Strassmann split a uranium nucleus apart in 1938.

Nuclear fission is the breaking apart of a large nucleus into two roughly equal nuclei. It turns out that to get a uranium nucleus to split into smaller pieces, one must hit it *gently* with a slow-moving neutron. Electrons are too light to do much damage, and protons or positively charged alpha particles are deflected by the large positive charge of the uranium nucleus and therefore can't get close enough to do any harm. Thus, until the discovery of the neutron by Chadwick in 1932, there was not a suitable tool with which to strike the uranium atom.

However, the neutrons released from radioactive decays in Chadwick's experiment were too energetic. A fast neutron has a large momentum, and through the de Broglie relationship (Chapter 3), the larger the momentum, the smaller the de Broglie wavelength. Finding the nucleus within an atom is always a difficult trick—if the electron's probability cloud, which denotes the "size" of the atom, were the size of your thumbnail (about one square centimeter) then the nucleus on the atom would be a single cell in the thumbnail. In 1937 Italian physicist Enrico Fermi discovered that passing a beam of neutrons through a length of wax caused the neutrons to slow down as they collided with the large paraffin molecules, but not come to rest, as they did when striking a similar length of lead. The slower the neutron is moving, the lower its momentum, and the larger its de Broglie wavelength. A larger wavelength gives the neutron more of a chance to intersect with the nucleus's matter-wave, just as you have a greater chance of coming across a bush in a garden at night if you walk with your arms outstretched rather than flat against your sides.

If the neutron strikes the uranium nucleus, then there is a chance that the strong force within the nucleus will capture this neutron (recall that the strong force has a very short range, and the neutron has to be right at the nucleus to feel it), making the uranium nucleus slightly heavier. But the tower of protons and neutrons in the uranium nucleus is already barely stable, and the addition of one more neutron turns out to be too much for the nucleus to support. So it usually tumbles into two smaller nuclei (typically krypton, with thirty-six protons and eighty-nine neutrons, and barium, with fifty-six protons and 144 neutrons, but alternative fracture products are observed), along with releasing either two or three more slowly moving neutrons,* and energy, in the form of kinetic energy of the smaller nuclei and gamma rays.

Where does the kinetic energy of the nuclear fission by-products come from? Electrostatics. While gravity and electromagnetism can exert a force even when objects are miles and miles apart (though the force gets weaker the greater the distance), the strong force holding the nucleus together disappears for lengths larger than the diameter of a neutron. Consequently, once the two large nuclear fragments break apart in the fissioning uranium, there is no strong force to hold them. But the thirty-six protons in the krypton nucleus and the fifty-six protons in the barium nucleus repel each other, and as they are initially very close, the repulsive force between them is strong. The kinetic energy of the nuclear fission products, which accounts for the horrible destructive capacity of an atomic blast, derives from basic electrostatics. Elements such as uranium or plutonium are easier to break apart than lighter elements, but *all* matter would violently explode if the strong force could be, even momentarily, turned off, as in *Watchmen*'s unfortunate Dr. Osterman from Chapter 5.

Many chemical reactions, such as when dynamite undergoes combustion, give off heat as a by-product. By "heat" I mean that the reaction products have a larger kinetic energy than the initial reactants. Nearly all chemical reactions have an energy scale of roughly one electron Volt per molecule, within a factor of ten or so (that is, sometimes the reaction takes a fraction of an electron Volt, while in some other cases, depending on the chemistry, the reaction could involve ten electron Volts or more). In contrast, a single uranium nucleus undergoing fission and splitting into two smaller

---

* Depending on the detailed decay fragments of the fissioning uranium nucleus. How exactly the unstable uranium nucleus decays into smaller nuclei is a complicated process.

nuclei releases about two hundred million electron Volts of energy. Consequently, the energy released in fission is much higher, per atom of initial material, than in a chemical reaction. But two hundred million electron Volts, from a single uranium atom, would be less noticeable than a mosquito bite. By gathering together several thousand trillion trillion uranium atoms, the resulting energy released can be devastating, even though these thousand trillion trillion uranium atoms would weigh only a few pounds. It would take more than twenty thousand tons of dynamite to release an equivalent amount of energy.

A given mass of uranium is dangerous, but half this mass is not. Why not? When the uranium nucleus captures a slow-moving neutron and fissions into two lighter nuclei, it also releases two or three slowly moving neutrons. Thus, the decay of one uranium atom provides the means to cause two more uranium nuclei to undergo fission, and each one of those can make two more nuclei decay. Starting with the fission of a single atom, a large number of additional atoms can be induced to decay in a chain reaction—but only if the neutrons emitted from the first uranium atom strike other nuclei. Remember that most of the atom is empty space and that the diameter of the nucleus is only one ten thousandth that of the atom itself. If the decaying uranium atom does not have a sufficient number of other atoms surrounding it, then there will be low-level decays that provide energy (useful for an electrical power plant) but not enough reactions to yield an explosive chain reaction.

The trick to making an atomic bomb is to have two separate pieces of uranium, each less than the "critical mass" (so defined as at this mass a chain reaction is ensured), and bring them together into one volume quickly enough that the reactions do not die out but continue to grow. It's not the mass itself that is critical for a chain reaction, but the number of uranium atoms, so that the released neutrons have a high probability of striking another nucleus and initiating another fission event. In this case a hundred pounds of uranium is transformed into an atomic bomb that can annihilate several square miles and cause extensive damage at larger distances.

Children in the early 1950s could learn all about radioactivity if their parents shelled out fifty bucks for the Gilbert's U-238 Atomic Energy Lab. This kit was the nuclear physics version of a chemistry set and came complete with radioactive sources that emitted alpha, beta, and gamma radiation, a Geiger counter, and a mini–cloud chamber for seeing the tracks created by high-speed radioactive particles. The kit included both an instruction manual and an informational comic titled *Learn How Dagwood Splits the Atom.* This comic featured text that was scientifically thorough and accurate, with an introduction by Joe Considine, an International News Service correspondent who covered the Bikini Atoll nuclear tests and wrote the script for the 1947 docudrama about atomic energy *The Beginning or the End* (not to be confused the 1957 science fiction film *The Beginning of the End,* which featured the attack of radioactive giant locusts), and a foreword by Lieutenant General Leslie R. Groves, the head of military operations at the Manhattan Project. In the accompanying comic, Mandrake the Magician shrinks Dagwood Bumstead, his wife, Blondie, and their kids and dogs to subatomic size, so that they, together with Popeye, Olive Oyl, and Wimpy, can observe firsthand the inner workings of nuclear decay and fission. Figure 22 shows a page from this booklet, as Dagwood, unable even with Popeye's assistance to overcome the strong nuclear force holding a uranium 235 nucleus together, is nevertheless able to initiate a chain reaction of fission decays when he uses a "neutron bazooka" to strike the nucleus just right.

While the world read in their newspapers on August 7, 1945, of the previous day's successful detonation of an atomic bomb by the U.S. military over Hiroshima, Japan—this was *not* the first time atomic weapons entered the public consciousness. Figure 23 shows a Buck Rogers newspaper strip published in 1929. When the submarine Buck and his colleagues are on is held fast by a giant octopus, their only hope is to blast themselves free, using the awful destructive potential of an *atomic* torpedo. A full sixteen years before the Manhattan Project, Phil Nowlan and Dick Calkins, creators of the "Buck Rogers, 2429 A.D." comic strip were confident that their readers would know that an atomic torpedo was a more powerful version of the regular underwater missile.

Moreover, according to adventure pulp magazines, Japan knew as well of the ability of atomic weapons to destroy a major city, six years before the U.S. bombing of Hiroshima and Nagasaki. In *Secret*

Figure 22: Page from Learn How Dagwood Splits the Atom *in which Mandrake the Magician, having shrunk Dagwood Bumstead and his family to subatomic size, narrates the mechanism of a uranium fission chain reaction, while Dagwood grabs his daughter and tries to quickly exit the nuclear pile.*

Service Operator No. 5, issue # 47, published in September 1939, it is the United States that is attacked by the invading troops of the "Yellow Vulture," a thinly disguised, racist version of the Japanese Empire. In a tale titled "Corpse Cavalry of the Yellow Vulture," the troops of the Yellow Vulture obliterate Washington, D.C., killing the president, Agent Q-6 (father to Operator no. 5), and most of the Washington establishment by using an atomic bomb.

One of the earliest recorded uses in fiction of "atomic" as a modifier to signify the enhanced lethality of a weapon is in a 1914 science fiction novel by H. G. Wells. In *The World Set Free*, Wells describes atomic bombs raining down with horrible destructive power and dropped from noiseless, atomic-powered airplanes.

How did the general population know about "atomic weapons"

Figure 23: Buck Rogers, in his daily syndicated newspaper strip in 1929, employs an "atomic torpedo" to devastating effect.

years *before* the Manhattan Project? It was thanks in part to the writings of Frederick Soddy, Ernest Rutherford's colleague in earlier studies of nuclear radioactivity. Soddy penned a series of popular science books, the best known of which, *The Interpretation of Radium: Being the Substance of Six Free Popular Experimental Lectures Delivered at the University of Glasgow*, was a best seller when published in 1909. It made quite an impression on Herbert George Wells, who incorporated the concept of atomic-based weapons weighing only a few pounds and releasing tremendous energy and lingering radiation damage into his novel *The World Set Free*. In Wells's novel, an atomic war between the nations of Europe and the United States leads to the formation of a proto–United Nations, where the surviving world leaders decide to form a new world order and establish a one-world government based upon the principles of socialism, rejecting capitalism, which was to blame for leading the nations into a nuclear confrontation.

This novel made a strong impression on one particular reader in 1932. Both Wells's vision of a one-world government run by socialistic principles and, equally important, his descriptions of horrific atomic weapons galvanized Hungarian physicist Leo Szilard. This fan of Wells was no ordinary reader—Szilard would, in 1933, be the first to conceive of a possible nuclear chain reaction (patenting the idea in 1934—four years before Hahn and Strassmann first split a uranium nucleus!). In 1939, Szilard wrote a letter to President Franklin Roosevelt, signed by Albert Einstein, urging the development of a nuclear weapons program, which became the Manhattan Project. Thus a popular science book by Soddy, written for a general audience, inspired an H. G. Wells science fiction novel suggesting the possibility of atomic weapons, and this novel in turn was directly responsible for the creation of actual atomic bombs. When publisher Hugo Gernsback launched his science fiction pulp magazine *Amazing Stories* in 1926, with a reprint of a story by Wells, it is doubtful that he realized how prophetic would be his magazine's motto: "Extravagant Fiction Today . . . Cold Fact Tomorrow."

CHAPTER TEN

Radioactive Man

The fates of Mickey Rooney and Tor Johnson in *The Atomic Kid* and *The Beast of Yucca Flats*, respectively, are of course ridiculous, unrealistic portrayals of the effects of exposure to radiation. By the mid-1950s, Doris Day's lighthearted song about the wonders of a Geiger counter would give way to darker implications regarding the effects of nuclear weapon testing.

Ten years after the use of atomic bombs at the end of World War II, science fiction films would clearly and unambiguously establish that the real risk of exposure to radioactive fallout is unchecked gigantism. James Whitmore and James Arness battled ants mutated to the size of helicopters by lingering radioactivity in the New Mexico desert in the 1954 Warner Bros. film *Them!* Exposure to an atomic testing site would similarly transform Lieutenant Colonel Glenn Manning into *The Amazing Colossal Man* (1955), who would return to wage the *War of the Colossal Beast* (1958); feasting on fruits containing radioactive isotopes would create giant locusts, signaling *The Beginning of the End* (1957); a diet of radioactively contaminated fish similarly causes an octopus to grow to fantastic size in *It Came from Beneath the Sea* (1955); and radiation in a swamp would provoke *The Attack of the Giant Leeches* (1959). Occasionally, radioactive exposure would instead lead to miniaturization, as reflected in the strange case of *The Incredible Shrinking Man* (1957) and the experiments of *Dr. Cyclops* (1940), whose shrinking beam was powered by atomic rays five years before the Manhattan Project.

"Radioactivity" is an umbrella term for particle or light emissions from nuclei. As discussed in the previous section, when electrons in an atom move from one quantized energy level to another, they do so via the emission or absorption of light,* which can span a wide range of wavelengths, from the microwave and infrared, to visible light, to ultraviolet and X-rays. Application of the rules of quantum mechanics to the protons and neutrons inside the atomic nucleus find that similarly, only certain quantized energy levels are possible. The energy spacing between these quantized levels is much larger than in the atom, thanks to the Heisenberg uncertainty principle. As the spatial extent of the nucleus is much smaller than that of the atom itself, the uncertainty in the location of the protons and neutrons is reduced. Consequently the uncertainty in the value of their momentum is increased, and the larger the momentum (mass times velocity), the greater the kinetic energy (momentum squared divided by twice the mass). While typical electronic transitions in an atom involve energies of about a few electron Volts, and occasionally one can observe X-ray emission, which has an energy of a thousand electron Volts, nuclear energy transitions involving electromagnetic radiation consist of gamma rays with energies of several million electron Volts.

As the protons and neutrons inside the nucleus settle from a higher energy level to a lower level (referred to as the "ground state"), there are other ways for them to shed energy aside from emitting gamma-ray photons. There are some nuclei that can lower their energy by emitting an alpha particle (consisting of two protons and two neutrons). The two protons and two neutrons that comprise a helium nucleus are very tightly bound to each other, so if the large, excited nucleus is going to emit any of its protons or neutrons, it is energetically favorable to do so in packets of alpha particles, rather than expending energy breaking the alpha apart. In this way the number of protons inside the larger nucleus decreases by two, so the electronic repulsion between the protons is reduced as well. The alphas come out with a considerable amount of ki-

---

* There is a process by which, when an electron drops from a high-energy state to a lower level, another electron in the atom is ejected. But for the most part electronic transitions within the atom involve emission or absorption of photons.

netic energy (several million electron Volts, typically). This made them convenient probes for Rutherford when studying the structure of the atom—investigations that led to the discovery of the nucleus.

Even though the nucleus can lower its energy by ejecting an alpha particle, the particles within the alpha are still subject to the strong force, which acts like a barrier holding the subatomic particles together within the nucleus. This barrier is high enough that ordinarily one would not expect any alpha particles to be able to leave the confines of the nucleus. Since alpha particles *have* been observed exiting the nucleus, there must be a mechanism by which they are able to leak out through this barrier. Here the bizarre phenomenon of quantum mechanical tunneling comes into play. The strong force is so effective at holding the nucleus together that the alpha particle has only one chance in one hundred trillion trillion trillion of escaping. However, its small spatial uncertainty within the nucleus leads to a large momentum uncertainty, and it "rattles around" inside the nucleus, striking the strong-force barrier a billion trillion times a second. Consequently, if one waits several billion years, one will see an alpha quantum mechanically tunnel outside of a nucleus. Once beyond the range of the strong force, the alpha particle is propelled at a high velocity by the same electrostatic repulsion that imparted energy to the fragments of a fissioning uranium nucleus.

Several billion years is a long time—so how are we able to see alphas emitted by radioactive isotopes without waiting so long? The answer to this question leads to an understanding of the concept of a radioactive half-life and in turn elucidates how we know the age of the Earth.

First a basic point about probability: In a lottery involving the random drawing of three digits from 000 to 999, there are one thousand possible outcomes. The lottery office draws the three digits at random, so one day the winning number may be 275 and the next it may be 130 or 477, and so on. If I purchase a ticket with one particular combination, say 927, there is thus one chance in a thousand that I will win the jackpot. Assume that I always play this same number, 927. I could win on the very first day. It's possible, though there is only one chance in a thousand that I will. It is conceivable that I may have to wait extremely long, much longer than a thousand draws, before my one ticket matches the three numbers. Certain combinations may appear as winning numbers many times before my particular ticket pays off.* I therefore may need to play the game for a long time before my ticket matches that day's winning numbers.

One important similarity between the lottery scenario and the decay of unstable nuclei is that for both, the chance of an "event" occurring (either matching your ticket's numbers with that day's drawings, or having the nucleus undergo a transition to a more stable configuration, with the release of radiation) is the same on any given day. In a real, standard lottery run by most states, there is no restriction on whether a given set of numbers (from the predetermined pool of possible numbers) can be repeated before all other possible combinations are drawn. On any given day, one particular combination of numbers is as likely as any other. Similarly, as the quantum mechanical transition to a lower energy configuration is a probabilistic occurrence, the nucleus is as likely to decay on the first day, the one hundredth, or the millionth. There is no upper limit on how long the nucleus can exist in the excited state before radiating back to a lower energy state. If the nucleus is able to remain in the excited state for a long time, it is not "due" or "expected" to undergo radioactive decay but is as likely to relax to the ground state on the millionth day as on the first. If one plays the lottery long enough, eventually every number that can occur will be drawn. Similarly, if one waits long enough, every unstable nucleus will decay to a lower energy state.

Depending on the nucleus and the nature of the unstable excited state it is in, the probability of decay may be very high or very low. In the lottery analogy, you may need to guess only one number from 0 to 9 in order to win the jackpot, or you may need to match seven random two-digit numbers in precise order. In the first case one would not need to play the game very long before winning, while in the second case it could take much longer than several lifetimes (if the lottery selected fresh numbers every day) before a

---

* My track record playing the lottery provides direct empirical evidence of this phenomenon.

winning match is obtained. Similarly, some elements' unstable nu-clei undergo radioactive decay within, on average, a few days or months, while others may take several billion years. However, in the first case there is no reason any given nucleus could not remain undecayed for a long time, while in the second situation there is no physical reason why any given nucleus could not decay almost immediately. It is possible to hit even a seven-digit lottery jackpot with your very first ticket, though I should be so lucky.

If I start with a large number of radioactive atoms, then a plot of the number that avoid decaying into some other isotope as a function of time follows what's termed an "exponential time de-pendence." To understand this concept, imagine a car driving at sixty miles per hour that suddenly slams on the brakes. How long does it take the car to come to a complete stop? If we assume that the brakes provide a constant deceleration of ten miles per hour per second, then in six seconds the car will come to a rest. What if the brakes provided a deceleration that depends on how fast the car is moving at any instant? That is, when the car is moving very fast the brakes provide a large force, slowing you down. But if you were driving much more slowly, in a parking lot, say, then the brakes would provide a lower force. If the deceleration is proportional to the velocity, then it turns out that the car never comes to a full stop! (Well, for long times it may be moving so slowly that we could for all intents and purposes say that it had stopped, but if we were to measure the speed, we might find that it is very, very small, less than one millionth of a mile per hour, for example, but never truly zero.) In the first case, that of a constant deceleration, the auto's speed decreases linearly with time. In the second situation, where the deceleration varies with the speed, initially the car slows down dramatically, as it is moving fast and that means the decel-eration is large. But as it goes slower and slower, the braking force decreases, so that for long times it is moving very slowly, but the brakes are exerting only a very weak force. A plot of the car's speed against time would be a concave curve called an "exponential decay function."

While the slowing automobile with velocity-sensitive brakes is artificial, the reverse phenomenon—exponential growth that leads to faster and faster increases—is more familiar, at least for those who have watched their savings grow through compound interest. A small amount deposited in the bank that earns a steady fixed interest rate, compounded continuously, will show a small increase initially. But as time progresses, both the original investment and the total interest earned will be subject to the same interest rate, and the returns will soon become much larger as your bank balance benefits from an exponential growth.

Just such an exponential dependence is found for the decay of tritium, an unstable isotope of hydrogen. Normally hydrogen has one proton in its nucleus. The neutrons, participating in the strong force, are needed in larger nuclei to overcome the electrical repulsion between protons. As hydrogen has only one proton in its nucleus, it is the only element that does not *need* neutrons, though it is possible for neutrons to be present in the hydrogen nucleus. In hydrogen, one electron is electrostatically bound in a quantum mechanical "orbit" to the single proton in the nucleus. As the chemical properties of an atom are determined by the number of electrons it possesses, which in turn are set by the number of protons in its nucleus, one could form an alternative form of hydrogen containing one proton and one electron, with an extra neutron in the nucleus, and it would behave, for the most part, like ordinary hydrogen. We would call this isotope deuterium. If there were two neutrons and one proton in the nucleus, about which one electron "orbits," this isotope is termed "tritium."*

As illustrated in Figure 24, tritium is unstable and, through a mechanism I describe in the next chapter, decays to form an isotope of helium, along with a high-speed electron (a beta ray) like those in Chapter 8 responsible for Dr. Manhattan's blue glow. Figure 24 shows another page from *Learn How Dagwood Splits the Atom*, whereby the addition of two neutrons to a hydrogen nucleus (that is, a single proton) yields an unstable result. One of the neutrons converts to a proton and another electron, through a mechanism governed by the weak nuclear force, discussed in detail in the next chapter. The decay rate of tritium is very fast, such that for a given nucleus, after only about twelve and a half years, there is a fifty-fifty chance of the isotope decaying.

If the decay rate is so fast, why is there any tritium still around?

---

* A nucleus with two neutrons and two protons is called helium, or, when ejected from a larger unstable nucleus, an alpha particle.

Figure 24: Page from Learn How Dagwood Splits the Atom *in which Dagwood, his son Junior, and his dog Daisy witness the radioactive transformation of a tritium nucleus into an isotope of helium.*

Because it is constantly being created, when high-speed neutrons formed from cosmic rays collide with nitrogen atoms in the atmosphere. The now unstable nitrogen nuclei decay to form normal carbon and tritium. The tritium generated in the upper atmosphere can be captured by oxygen atoms and forms a version of "heavy water" (remember that aside from the heavier nucleus, tritium behaves chemically the same as normal hydrogen). This tritium-rich water reaches the ground in the form of raindrops. Because we know the decay curve of tritium, comparisons of water from the surface of the ocean to that obtained from greater depths enable determinations of the cycling time for oceanic circulation currents.

Ideally, in order to measure the time dependence of the tritium decay, one would like to have samples of rainwater from more than a hundred years ago, as well as more recent years all the way to the present. By measuring the fraction of tritium as a function of the age of the water, one could verify the exponential time dependence of its decay. The problem is that one does not have bottles of rainwater dating back more than a century. In a 1954 paper in the *Physical Review*, Sheldon Kauffman and Willard F. Libby did the next best thing and examined the tritium content of vintage wines. As shown in Figure 25, a plot of the tritium concentration per wine bottle as a function of time, determined from the vintage label, shows that, when measured in 1954, the tritium concentration was very high in a 1951 Hermitage Rhone, but the concentration was dramatically lower in a 1928 Chateaux Laujac Bordeaux. The full curve is very well described by an exponential time dependence. Based on this curve, if in 1954 we wanted a wine with a tritium concentration half as large as that in the 1951 Hermitage, we would decant a 1939 vintage, from which we conclude that the "half-life" of tritium is 12.5 years.

Different radioactive nuclei have different decay rates. All unstable nuclei have exponential decay functions, but the time scale over which the decay occurs may be very different—from minutes to billions of years. Measurements of nuclei with short decay times, such as the tritium in wine bottles example, confirm that the number of nuclei that decay does indeed follow an exponential time dependence. The physics of the nucleus does not change depending on which element we are considering. For those nuclei that have

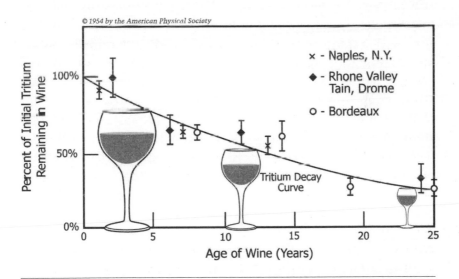

Figure 25: Plot of the time dependence of tritium concentration in "heavy water" contained in wine bottles. The age of the water sample is determined by the vintage printed on the bottle's label. The longer one waits, the less tritium is present, due to radioactive decay. The solid line is a fit to the data of an exponential time dependence, with a half-life of 12.5 years. Reprinted figure with permission from S. Kaufman and W.F. Libby, Physical Review 93, 1337 (1954).

very low decay rates, so that the time to decay is very long, we can nevertheless measure the initial portion of the exponential decay. Mathematical fitting of this curve indicates when the decay function is expected to reach the 50 percent point, and thus we can determine that the half-life of uranium, for example, is several billion years, even though we have not sat in the lab for this length of time to measure the full decay curve.

For a radioactive nucleus with a half-life of one year, if I start with a million atoms, after one year I will have approximately half a million remaining (there will typically be fluctuations about this average number of half a million, as the decays are probabilistic). As the decay rate is independent of the age of the atom, then in the next year, 50 percent of the remaining atoms will decay. That is from an initial number of one million, I will have approximately half a million after one year, a quarter of a million after two years, 125,000 after three years, and so on.

Because the time necessary for one half of the initial population of nuclei to decay is precisely known, we can use carbon dating to determine the age of archeological artifacts. Let's say we start with a million unstable isotopes of carbon. Normally carbon has six protons (and a corresponding six electrons in quantum mechanical "orbits") and six neutrons in its nucleus and is as stable as anything we know of. As there are twelve particles in its nucleus, this form of carbon is called carbon 12. Occasionally collisions with cosmic rays lead to the creation (through a process that we don't have to worry about now) of a form of carbon with six protons but eight neutrons in its nucleus. As it has the same number of protons and electrons as carbon 12, this heavier isotope is chemically identical to normal carbon. However, this form of carbon with eight neutrons (called carbon 14) is unstable and beta decays into nitrogen 14.

The rare heavier carbon 14 is constantly being created by random collisions with cosmic rays and is constantly decaying away into another element. A very small but constant percentage of the carbon in the world is heavy, unstable carbon 14. This holds for the food we eat, the clothes we wear, and pretty much everything that contains carbon atoms. Consequently, a small fraction of the carbon in our bodies is this unstable heavier form. The half-life

for heavy carbon to decay is about 5,700 years. So normally we ingest heavy carbon by its random presence in the food we eat, and
we lose heavy carbon through normal biological processes when we
eliminate old cell material. This process comes to a rather abrupt
stop when we die (the flux of cosmic rays on the Earth's surface is
low enough that we don't have to worry about carbon 14 creation
in our corpse). At death the amount of heavy carbon in our bodies,
our skin, our tissues, and our bones is fixed. A future archaeologist,
finding our skeletons, measures the quantity of carbon 14 and finds
that it is only half of the normal amount of carbon 14. She can then
confidently state that we died approximately 5,700 years ago. If the
amount of heavy carbon is one quarter of the current level of carbon 14, then two half-lives must have passed, and our death is
placed at roughly 11,400 years in the past. In this way any material
containing organic matter, whether it be ancient bones or the
shroud of Turin, can be dated from its last point of carbon intake.
Willard Libby, who used old wine to obtain new measurements of
tritium decays (Figure 25) shared the 1960 Nobel Prize in chemistry
for developing carbon 14 dating.

Longer-lived isotopes, such as uranium 235 and uranium 238,
have half-lives of roughly billions of years. These two forms of
uranium were generated in a supernova explosion that created all
the atoms that went on to form the planets and moons in the solar
system (more on this later). Assuming that initially they are created in equal concentrations, ascertaining their half-lives through
independent measurements, and seeing the fraction of uranium
235 and uranium 238 present on the Earth today, we can calculate
how long the Earth has been around to give the uranium isotopes
a chance to decay to their present proportions.* The answer turns
out to be about 4.5 billion years.

We thus know the age of the Earth through our understanding
of quantum mechanics, the same quantum physics that underlies
the field of solid-state physics. Without quantum mechanics, there
would be no semiconductor revolution, and the nearly countless
electronic devices we employ would not be possible. It is of course

---

* The full analysis involves studies of meteorites, and is more complicated than summarized here.

your right to believe that the Earth is actually much younger than its age determined by radioactive isotope dating, but to be consistent, you should stop believing in your cell phone, too!

Elements that emit gamma rays, alpha particles, or beta particles are *radioactive*—while materials that are exposed to these nuclear ejections are described as *irradiated*. As mentioned at the start of this chapter, science fiction films in the 1950s ascribed to irradiation the mutation of animals and people into giants, though occasionally a miniaturization effect was possible. What exactly are the real, non–Hollywood movie, effects of exposure to radiation? Not all radioactivity is created equal, and some is more harmful than others.

The emission of radioactivity results when a nucleus makes a quantum transition from a high energy state to a lower energy configuration. Recall that the energy spacing between quantum states in the nucleus is on the order of a million electron Volts, while electronic states in an atom are on the order of a few electron Volts. Electronic transitions involve energies in the ballpark of visible light, while the energy scale of nuclear quantum jumps is much larger. When the electrons in a neon atom make quantum transitions, they emit red light, which we associate with neon signs. When the neon atom's nucleus makes a transition, the energy is about a million times greater and has the potential to do extensive damage. We evolved in a sea of visible and ultraviolet light, and aside from a sunburn (and the concomitant long-term skin damage), this radiation does not harm us. Light a million times more energetic is rare, and we are not equipped to shrug off such radiation.

If you were wandering around a nuclear weapon testing site, you would be exposed to the fallout—radioactive isotopes that are created as the by-products of the fission reaction. A variety of secondary, hazardous unstable nuclei can be generated depending on the nature of the atomic blast. They, in turn, emit radioactivity as they relax to lower energy states. You would wish that the radioactivity present would be alpha particles, consisting of two protons and two neutrons, rather than beta rays, or high-speed electrons. The energy, in the form of kinetic energy, of either the alpha particle or beta ray is about a million electron volts. The mathe-

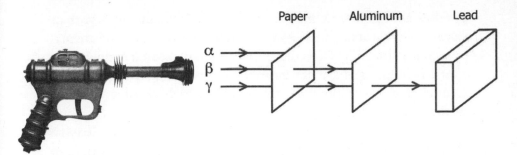

Figure 26: *A sketch of a 1936 Buck Rogers toy ray gun and nuclear radioactivity consisting of alpha particles, beta rays, and gamma ray photons. (The toy shown is a Buck Rogers Disintegrator Pistol, model XZ-38, and did not actually emit high-energy sub-atomic particles). All three types of radiation have roughly the same energy of a few million electron Volts, but they are stopped by different levels of shielding. The alphas are blocked by a sheet of paper, which the betas can penetrate, but are themselves stopped by a thin aluminum sheet. High-energy light in the form of gamma rays passes through both, and a thick block of lead is needed to stop them.*

matical expression for kinetic energy is KE = (1/2) mv², where m is the mass of the object and v is its velocity. Therefore, for a given kinetic energy, the larger the mass, the smaller the velocity. Protons and neutrons have a mass roughly two thousand times larger than that of an electron, so the alpha particle is nearly eight thousand times more massive than an electron. Thus, if both an alpha particle and a beta particle have comparable energies, as they both arise from nuclear quantum transitions, the alpha will be moving nearly ninety times slower than the beta. The slower a particle moves through matter, the more time it spends near each of the atoms in the object, increasing the likelihood of losing energy through collisions with the electrons surrounding each atom. Slower alpha particles can be stopped by a single sheet of paper, and they almost never penetrates a person's clothing, while it takes a quarter-inch-thick sheet of aluminum to stop much faster beta rays, and they can indeed get under your skin.

Gamma-ray photons are high-energy light, several hundred times more energetic than X-rays. As gamma rays are photons of light and uncharged, they do not interact directly with the electronic charges in atoms, which makes them much harder to stop.

It takes about one-half inch to an inch of dense material such as lead to stop gammas, and they can penetrate through an entire person. Some nuclei (such as uranium 238) may also emit neutrons* that in themselves are not harmful, but when they collide with hydrogen atoms in the body, the resulting high-speed ricocheting protons can be damaging.

If you are unfortunate enough to ingest an unstable nucleus, so that it is inside you when it undergoes radioactive decay, then even alpha particles can be deadly. Rather than striking the dead skin cells on your epidermis, which you slough off naturally, alpha particles inside you have a direct path to your internal organs. In this case the alpha particles prove to be very efficient in stripping electrons from the atoms they strike, disrupting the chemical bonds within the cell and causing extensive chromosomal damage.

In 2006 Russian journalist Alexander Litvinenko was murdered when he drank tea that was spiked with polonium 210. This unstable nucleus has a half-life of just over 138 days and emits high-energy alpha particles when it decays. A pound of polonium 210 releases energy at the rate of nearly 64,000 Watts. Because its probability of radioactive decay is so high, even 0.05 micrograms of polonium 210 is considered to be lethal (it is believed that Litvinenko had 10 micrograms in his body at the time of death). So, alphas on the outside, not too much of a problem—on the inside, a rather big problem. Which is probably why it was fortunate that Gilbert stopped marketing their U-238 Atomic Energy Lab in 1952. I mentioned in the last chapter that this kit contained a mini–cloud chamber. Part of the radioactive elements supplied with this chamber was a small piece of potent, though short-lived, radioactive polonium 210!

Of course, the fact that some forms of radiation can cause sufficient damage to kill the living cells they impinge can be a good thing. The cells in a piece of meat from the butcher are nonliving, but the bacteria within the meat that can cause salmonella or other diseases are very much alive. Exposure to radiation does no significant damage to the cells in the food that are already dead but

---

* By the time neutrons were discovered, the identity of nuclear radiation was better understood, and the "ray" nomenclature was no longer employed.

can penetrate and kill the bacteria living in the food, thus making the food much safer to eat. The exposure to radiation does not convert the stable nuclei in the food's atoms into unstable nuclei, and they will not in turn emit their own radioactivity. A material that has been irradiated does not (with very few exceptions) itself become radioactive.

A lot of the harm of nuclear radiation is caused when either negatively charged beta rays or positively charged alphas or gamma-ray photons collide with atoms and cause them to lose their electrons. This process is called "ionization," and when an atom loses some or all of its electrons, its chemical properties can be radically changed. Sometimes these changes are beneficial. The Earth is constantly being bombarded with cosmic rays, which are primarily (though not exclusively) high-energy protons that come from sources as close as our sun and as distant as other galaxies. When these protons (a few of which have energies of a million trillion electron Volts) strike the atmosphere, they can generate a slew of other elementary particles moving near the speed of light. When some of these particles strike the DNA in our cells, they can cause ionizing damage and alter the chemical properties of our genetic code. If the affected DNA is in a sperm or egg cell, these changes may be passed along to offspring. In this way exposure to cosmic rays is a natural source of genetic mutation, leading to biological modifications that can be harmful but occasionally improve an organism's fitness to its environment.

But beneficial mutations that do not harm the original organism and lead to genetic alterations that improve the offspring's reproductive success are extremely rare. More commonly, chemical modifications induced by ionizing radiation can destroy cells themselves or induce deleterious alterations in chromosomes or DNA. This damage often leads to the formation of malignant cancerous tumors, quite different from the runaway cell growth presented in science fiction movies such as *The Amazing Colossal Man* or *The Attack of the Giant Leeches*.

CHAPTER ELEVEN

Man of the Atom

Before there was physicist Jon Osterman, there was physicist Philip Solar. In the 1986 DC comic book *Watchmen*, Osterman was disintegrated by the accidental removal of his intrinsic field at Gila Flats and reconstructed himself as the super-powered Dr. Manhattan. In the 1962 Gold Key comic book series *Solar—Man of the Atom*, Phillip Solar was exposed to a lethal dose of radiation in a sabotaged nuclear research experiment at Atom City, yet survived, though he acquired "quantum powers." In issue # 2 he is vaporized by an atomic bomb blast but manages through sheer force of will to reconstitute himself as, well, Dr. Solar, which was his name after all. As a survivor of graduate school myself, I can empathize with Osterman and Solar's inclination to retain the title associated with their Ph.D.'s, in the lab or as a superpowered hero. Once you've passed through the crucible of a graduate school candidacy exam, having to reassemble yourself up from the sub-atomic level is not as challenging as you might think.

In writer Alan Moore's initial outline of the DC comic book miniseries *Watchmen*, he intended to use comic book characters created by Charlton, another comic book publisher. Charlton had declared bankruptcy, and the company had been acquired by DC Comics, home of Superman and Batman. Moore's initial outline for *Watchmen* made direct use of the Charlton characters, but the editors at DC Comics, seeing that some of these characters would not make it out of the miniseries unscathed, instructed Moore to instead employ alternate versions of the Charlton heroes. Dr. Manhattan is the analog of Captain Atom, an air force captain, who was disintegrated and (I'm sure you can see this coming at this stage) through force of will was able reassemble himself into a quantum-powered superbeing.

Captain Atom's powers were quantum based only in that he was able to manipulate energy, which he employed primarily for flight, superstrength, and energy blasts. Dr. Solar, though not a Charlton character, seems to be a closer antecedent for Dr. Manhattan, as Solar was also able to change size (*Dr. Solar* # 10 and # 11), split himself into multiple copies of himself (*Dr. Solar* # 12), and manipulate matter and energy, though unlike the blue Dr. Manhattan, Dr. Solar's skin turned green when he used his powers. There are just enough differences among Captain Atom, Dr. Solar, and Dr. Manhattan that it is unlikely that they are all the same person, on three different versions of Hugh Everett's many worlds, though further study appears warranted.

One of the more accurate manifestations of quantum mechanical powers was presented in "Solar's Midas Touch," in 1965's *Dr. Solar, Man of the Atom* # 14. In this tale an underwater nuclear reactor pile went critical when one of the control rods (whose role is to absorb neutrons, decreasing the rate of uranium fission, as described in Chapter 9) broke. Dr. Solar, whose powers are normally energized by exposure to radiation, went underwater to fix the reactor but found himself weakened by the reactor's radioactivity (through a process not clearly explained in the comic). Eventually he was rescued by a worker wearing a lead-lined safety suit, who would have done the job in the first place if Dr. Solar hadn't attempted to "save the day." The additional radiation he absorbed from the reactor temporarily endowed Solar with a new superpower. As illustrated in Figure 27, whenever Dr. Solar comes into physical contact with an object, he transmutes it into the next element up the periodic table. In Figure 27, he transforms gold, with seventy-nine protons, into mercury, with eighty protons; earlier he grasps a copper rod (twenty-nine protons) and converts it into zinc (atomic number 30); and even when flying he begins to choke when the oxygen (atomic number 8) turns into fluorine gas (with nine protons). This newfound power of Dr. Solar's appears to be the abil-

Figure 27: *In* Dr. Solar, Man of the Atom # 14, *an additional nuclear accident endows Dr. Philip Solar (wearing the scuba suit and visor) with the temporary ability to induce beta decay via the weak nuclear force in any object he comes into direct contact with, thus transmuting gold into mercury.*

ity to initiate beta decay of the neutrons in any object he touches, inducing elemental transmutation via the weak nuclear force, an aspect of Watchmen's "intrinsic field" that we have not discussed much yet.

We saw in Chapter 9 that neutrons, through the strong force, hold the nucleus together by binding with protons and other neutrons and overwhelming the electrostatic repulsion that would, in their absence, cause the protons to fly out of the nucleus. Protons also exhibit the strong force, but without neutrons there is not sufficient binding energy to hold together a nucleus consisting only of protons. Neutrons themselves are not stable outside of a nucleus. A neutron sitting alone in the lab will decay into a proton and an electron with a half-life of about ten and one quarter minutes. The electron will be moving very near the speed of light, and when this process occurs within a nucleus, it is the source of the beta rays emitted from unstable isotopes.

As the total mass and energy of an isolated system must remain unchanged in any process, a "stationary" neutron* can decay only to fundamental particles with less mass than the neutron's. A neutron will thus decay into a proton, which has a slightly smaller mass, while a "stationary" proton could not decay into a heavier neutron. However, as the neutron is electrically neutral, and the proton is positively charged, the decay must also generate a negatively charged electron, in order for the total electrical charge to remain unchanged before and after the decay (we have not needed to invoke this principle before now, but another conservation principle in physics, comparable to conservation of energy or conservation of angular momentum, is conservation of charge, in that the net electrical charge can be neither created nor destroyed in any process). An electron is nearly two thousand times lighter than a proton, less than the mass difference between neutrons and protons, so adding an electron to the decay is still consistent with mass conservation. While a neutron decaying into a proton and an electron means that mass and electrical charge are balanced during the decay, examination of the kinetic energy of the proton and the high-speed electron (that is, the beta ray) and comparison to the rest-mass energy of the neutron indicates that some energy went missing in the process—not a lot, but enough to notice, and enough to cause trouble.

When physicists in the late 1920s discovered this phenomenon and realized that it appeared to violate the principle of conservation of energy, they were faced with two choices: (1) either abandon conservation of energy, at least for neutron decay processes, or (2) invent a miracle particle that was undetectable by instrumentation of the time but that carried off the missing energy. In 1930, Wolfgang Pauli (whose exclusion principle I address in the next section) suggested going with option 2. Knowing that this ghost particle had to be electrically neutral and had to have very little or no mass, Enrico Fermi called it the "little neutral one" in Italian, or "neutrino."[†]

---

* That is, one that is sitting still, isolated in space or moving but stationary to us if we moved along in the same speed and direction as the neutron (called the neutron's "rest frame").

† For technical reasons that need not concern us, when neutrons decay they emit a proton, an electron, and an antineutrino (the antimatter version of a neutrino). As I say, we need not worry about this particular detail here.

Detectors were eventually constructed to observe these particles, and their existence was confirmed in 1956. These particles not only really exist, aside from photons they are the most common particle in the universe. Their interactions with matter are governed by the weak nuclear force, which is one hundred billion times weaker than electromagnetism (the force by which electrons interact with matter). Neutrinos consequently barely notice normal matter (it takes more than two light years of lead—that is, a length of more than ten trillion miles—to stop one). If you hold out your thumb and blink, during that time period more than a billion neutrinos will pass through your thumbnail.

Dr. Solar, after his radiation overdose, must have gained an uncontrolled ability to induce beta decay in any object with which he came into contact. If a gold atom, with seventy-nine protons, seventy-nine electrons, and 118 neutrons, has one of its neutrons spontaneously decay into a proton and an electron, then it will have eighty protons, eighty electrons, and 117 neutrons. The lightest, stable configuration of mercury has eighty protons, eighty electrons, and 118 neutrons, so Dr. Solar will have created an unstable isotope of mercury in Figure 27. The half-life of this isotope of mercury with 117 neutrons is roughly two and a half days, so there will be time for Solar to finish his adventure and try to restore the transmuted mercury back to its original golden state. While transforming one element into its periodic-table neighbor via neutron beta decay is not quite the alchemist's dream of transmuting lead into gold (normal beta decay would convert platinum, with seventy-eight protons, into gold, with seventy-nine protons, so, depending on world exchange prices, you may wind up losing money on the deal), a process known as "reverse beta decay" would turn mercury into gold. While we cannot initiate such a conversion on Earth at will, fortunately this inverse process occurs constantly in the center of the sun, keeping the sun shining and providing the basis of all life.

The light from the sun—which is transformed by photosynthesis into chemical energy stored within plants, which in turn provides us with the energy we need to maintain our metabolisms—originates from nuclear transformations in the star's core. Four protons, that is, hydrogen nuclei, subjected to the extreme pressures and temperatures at the center of the sun, are fused together to form helium nuclei. But a helium nucleus consists of two protons and two neutrons, not four protons. Recall that neutrons are necessary as mediators of the strong nuclear force that holds the nucleus together. Thus, to make helium out of hydrogen, you first have to combine two protons and then through *reverse* beta decay convert one of the protons into a neutron.

I argued above that a single proton cannot convert into a neutron, as the mass of the proton is less than that of the neutron, and lighter objects cannot decay into heavier products. If two protons collide, the weak force operates on the protons, turning one into a neutron through reverse beta decay, as illustrated in Figure 28. The proton and neutron, subject to the strong force, become bound (now a deuterium nucleus—an isotope of hydrogen) and lower their en-

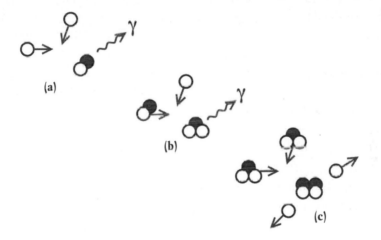

Figure 28: Sketch of the nuclear reactions in the center of the sun by which protons (hydrogen nuclei) combine to form alpha particles (helium nuclei). In step (a), two protons (represented by open circles) tunnel together, where the weak force converts one proton (open circle) into a neutron (dark circle). The proton and neutron then form a bound deuterium nucleus, with the release of a gamma ray photon (the positron and neutrino released are not shown for simplicity). The deuterium can then collide with another proton in step (b) and form a bound proton-proton-neutron nucleus, termed helium-3. In step (c) we indicate a possible reaction where two helium-3 nuclei collide and form a stable helium-4 nucleus (two protons and two neutrons), with the release of two protons and another gamma ray. Similar mechanisms result in the fusion of helium nuclei to synthesize heavier elements, such as carbon and oxygen, and up.

ergy compared to an isolated proton and neutron. This lower energy is reflected in a smaller mass for the deuterium nucleus, relative to a free proton and neutron. While the mass difference is very small, through $E = mc^2$ the energy difference of the bound deuterium is significant, and it emits a 2.225-million-electron-Volt gamma-ray photon during formation. In addition to the neutron generated by the weak force, the reaction creating a deuterium nucleus yields an antimatter electron (which has a positive electrical charge like a proton, but the mass of an electron) and a neutrino.

The weak force extends over a length scale roughly one thousand times smaller than that of the strong force, which itself acts only over distances less than the diameter of a nucleus. Two protons, both being positively charged, repel each other, and the closer they are, the greater the repulsive force. So one must force the two protons very close together, overcoming their electrical repulsion, in order for there to be an opportunity for the weak force to transform, through reverse beta decay, one of the protons into a neutron. The temperatures and pressures in the center of the sun are enormous, so that there are many opportunities for high-velocity collisions between two protons. However, even at the center of the sun the proton speeds are not sufficient to overcome the electrical repulsion when they draw too close. How do they manage to get past this electrical barrier? Through quantum mechanical tunneling!* Just as the alpha particles in radioactive decay use tunneling to escape the strong-force barrier around the nucleus that keeps the protons and neutrons together, the two protons that join together, forming the simplest isotope of hydrogen, must tunnel to overcome the barrier of their mutual repulsion.

The deuterium nucleus created in the center of the sun is stable and continues to collide with other protons. Combining this deuterium with another proton forms a nucleus with two protons (that is, helium) but only one neutron (making it helium 3, a lighter

---

* Famed astrophysicist Sir Arthur Eddington was one of the first to recognize that the sun's energy resulted from nuclear reactions. However, prior to the development of quantum mechanics, many physicists objected that the temperatures at the center of the sun were too low to enable such processes. To this Eddington replied in 1920, "We do not argue with the critic who urges that the stars are not hot enough for this process; we tell him to go and find a hotter place."

isotope of helium). Here again quantum-mechanical tunneling is required to get the second proton close enough to the deuterium nucleus, overcoming the electrical proton-proton repulsion, for the strong force to hold the second proton in the now larger nucleus. The lower energy of this bound state results in the release of another gamma-ray photon. This reaction is much more likely than for two deuterium nuclei to combine to form normal helium (two protons and two neutrons).

There are then many different ways that the helium 3 or deuterium nuclei can interact to form a stable helium nucleus, all of which involve quantum mechanical tunneling to get the positively charged nuclei close enough for the strong force to operate, resulting in the release of a great deal of energy in the form of kinetic energy of the nuclei, gamma rays, and neutrinos. The neutrinos pass right through the sun and head off in all directions, while the gammas heat up the nuclei and electrons in the center, accelerating them and causing them to emit electromagnetic radiation at all wavelengths. The light created in the center of the sun is scattered many, many times before reaching the surface, where it then takes the brief, eight-and-a-half-minute journey to Earth. Before reaching the surface, the average photon spends forty thousand years colliding with the dense nuclear matter in the sun's interior. The outward energy pressure counteracts the inward gravitational pull and keeps the diameter of the sun fairly stable.

In addition to providing us with energy, this fusion process is the mechanism by which elements heavier than helium are synthesized. Our sun is actually a second-generation star that formed after a much larger star passed through its life cycle and "went supernova." Our sun converts a great deal of hydrogen as it generates energy—approximately six hundred million tons per second. But eventually stars exhaust their supply of hydrogen, and the star collapses until the temperature and pressure rise to the point where helium nuclei begin to fuse, forming carbon. The process continues, generating nitrogen, oxygen, silicon, and other heavy elements up the periodic table to iron and nickel. However, the larger the nucleus created, the less energy is released per reactant, and at the iron/nickel point, the outward flow of energy is insufficient to counteract the inward gravitational pull. At this stage the star collapses onto itself; in the process, all elements heavier than iron are created, and there is an explosive outpouring of energy as the star becomes a supernova, releasing as much energy in a period of several weeks as our sun does over its entire lifetime. It is from the elements synthesized in a much larger star that lived and underwent a violent demise that the planets and sun of our solar system formed.

The power of the atomic bomb results from the breaking apart of large nuclei, such as uranium or plutonium, in a fission process, described in Chapter 9. Current nuclear power plants, such as the one that went critical and injured Dr. Solar at the start of this chapter, are fission reactors. They require rare radioactive isotopes as fuel, and their by-products are unstable isotopes, which are themselves radioactive and harmful to people. After the atomic bomb, the hydrogen bomb was developed. This weapon utilizes a fission reaction to initiate a fusion reaction—the energy of an atomic bomb is employed to force heavy isotopes of hydrogen and helium to fuse and release even more energy. For more than fifty years, scientists have been attempting to construct a fusion reactor that could create energy for electricity production, harnessing the power of the hydrogen bomb and the sun for peaceful, controlled terrestrial needs. The required fuel for a fusion reactor involves isotopes of hydrogen (typically deuterium and tritium), which may be harvested from naturally occurring isotopes of seawater, and the reaction products are nonradioactive. The obstacle is to replicate, in a controlled manner, the temperatures and pressures at the center of the sun. While the engineering challenges have indeed been formidable, a consortium of nations including Europe, Russia, Japan, and the United States are constructing a pilot fusion power plant (the International Thermonuclear Experimental Reactor, or ITER) to examine the feasibility of using nuclear fusion for electricity generation.

Back in the late 1980s there was a brief flurry of interest in reports that nuclear fusion had been achieved in a small tabletop experiment involving the electrolysis of heavy water using a palladium electrode. This so-called cold fusion process proposed that the deuterium nuclei, embedded within the metal electrode, were undergoing fusion and creating helium nuclei, with a concurrent release of excess heat. Whatever was going on in their device, it was not nuclear fusion, and it's a good thing for the chemists involved in this project that they were not in fact generating fusion reactions. A by-product of this particular fusion reaction is high-energy neutrons that would have killed anyone unlucky enough to be in the lab at the time. Moreover, as discussed earlier, fusion reactions within the center of the sun, at temperatures of millions of degrees, require quantum mechanical tunneling for the protons to overcome their electrical repulsion. Fusion at room temperature in a palladium electrode is even more dependent on tunneling to proceed. A well established feature of quantum mechanics is that the tunneling probability is very sensitive to the mass of the object involved. The smaller the mass, the lower the momentum and the longer the de Broglie wavelength, which can extend farther through the forbidden region, increasing the probability of finding the object on the other side of the barrier. Yet the initial investigators of "cold fusion" found no difference whether they used heavy water or ordinary tap water, whereas the difference in mass should have had a large effect on the fusion process.

For cold fusion to be a real phenomenon, it would require a suspension or violation of the principles of quantum mechanics, which underlies our understanding of solid-state physics, lasers, transistors, and all of the personal electronic devices they enable. Nevertheless, one might be tempted to give these up, if we could make cold fusion a physical reality. After all, a small cylinder capable of generating the power of the sun would make an awesome power supply for a jet pack!

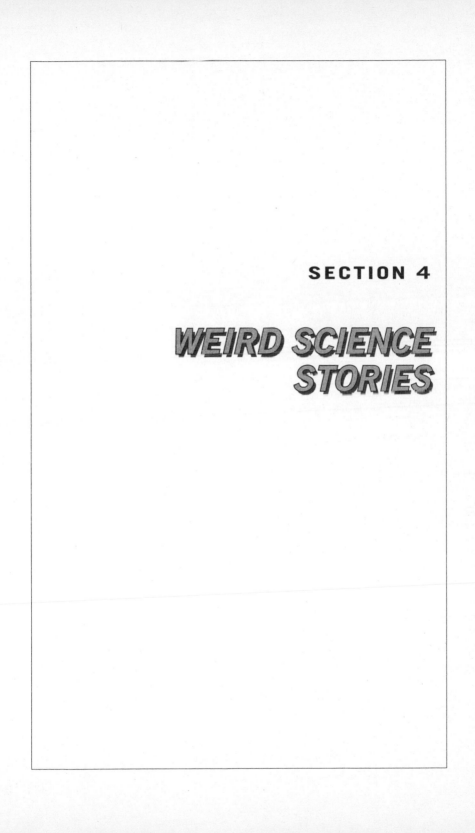

SECTION 4

WEIRD SCIENCE STORIES

CHAPTER TWELVE

Every Man for Himself

The agreement between theoretical predictions of atomic properties using quantum mechanics, such as the wavelengths of light emitted when an excited hydrogen atom relaxes back to its ground state, and experimental measurements of these wavelengths is nothing short of amazing. But if that were all that quantum mechanics could do, it most certainly would not have "made the future." We would still be living in the "vacuum-tube age" and would not have laptop computers, cell phones, DVDs, or magnetic resonance imaging devices.

The quantum descriptions of Schrödinger and Heisenberg accurately account for the properties of a single atom, but very rarely does one encounter an isolated, single hydrogen atom, or any type of atom or molecule by itself. A typical cubic centimeter of a liquid or solid, about the size of a sugar cube, contains roughly a trillion trillion atoms. The power of quantum mechanics is that it also provides an understanding of the properties of these trillion trillion atoms and accounts for why some materials are metals, some are insulators, and others are semiconductors. Fortunately for us, it turns out that if one understands the behavior when *two* entities are brought close enough to each other that their Schrödinger wave functions overlap, then this tells us nearly all we need to know to understand the results of a trillion trillion entities in close quarters.

Up to now, we have made extensive use of the first two quantum principles listed in Section 1: that light consists of discrete packets of energy termed photons, and that there is a wave associated with the motion of all matter. We have not needed to invoke the third principle: that all matter and light has an internal rotation that corresponds to discrete values of angular momentum. We would have needed this principle in order to understand details of how the electronic energy states in an atom are arranged, but for our purposes there was no call to head into this set of weeds. However, we cannot avoid a certain amount of weediness now, not if we wish to understand the basis for the semiconductor age and the foundations of the upcoming nanotechnology revolution.

In Chapter 4 we discussed the intrinsic angular momentum that all subatomic particles possess, termed spin. Associated with this internal spin is a magnetic field, so that every electron, proton, and neutron can also be considered a tiny bar magnet, with a north and a south pole (Figure 10b). While the concept of spin was introduced to account for experimental observations indicating that electrons possessed a built-in magnetic field, one cannot ascribe this magnetic field to a literal, physical rotation of the subatomic particles as if they were ballerinas. It is indeed confusing to imagine an intrinsic angular momentum, as integral to the properties of the electron and as real as its charge or mass, that is not associated with a literal rotation. Nevertheless, spin is the term that has stuck, and we adhere to this nomenclature, as we are nothing if not slaves to convention.

As mentioned in Chapter 4, the intrinsic angular momentum of electrons is exactly $\hbar/2$ (recall that $\hbar$ is defined as $h/2\pi$).* The "spinning" electron can have an intrinsic angular momentum of either $+\hbar/2$ or $-\hbar/2$, just as a real spinning ballerina can twirl either clockwise or counterclockwise. No other intrinsic angular momentum values are possible for electrons (or protons or neutrons). The collective behavior of quantum particles that have a spin of $\pm\ \hbar/2$ was first worked out by Enrico Fermi and Paul Dirac in the 1920s. In honor of their contribution, physicists refer to all spin $\hbar/2$ particles as obeying "Fermi-Dirac statistics," or by the shorter nickname of fermions. (Fermi got the sweet part of this deal—the

---

* Measured relative to a particular axis of rotation.

fact that electrons are spin $\hbar/2$ particles, and thus are fermions, has led to a host of quantities in solid-state physics as being labeled with his name—Fermi Energies, Fermi Surfaces, and so on—even though he made few direct contributions to this field of physics.)

Consider two fermions, such as electrons. It really is true that all electrons look alike. This is not the prejudiced opinion of an anti-Fermite, but a reflection of the fact that all fundamental particles of a given type are identical. There is no way to distinguish or differentiate between electrons, for example. Similarly, all protons are identical, as are all neutrons. These three subatomic particles have different masses and electrical charges, so they can be distinguished from one another. But if we bring two electrons so close to each other that their de Broglie waves overlap, then no observable property can possibly depend on which electron is which.

If I toss a rock into a pond, a series of concentric circular ripples forms (Figure 29a). When I toss two rocks into the water a small distance apart, each forms its own set of ripples, and the combined effect is a complicated interference pattern (Figure 29b). At some points the ripples from each rock add up coherently and create a larger disturbance on the water's surface than generated by each rock separately. At other locations the two ripples are exactly out of phase, so that one ripple is at a peak while the other stone's wave is at a trough, and the two exactly cancel each other out. Taken together, the resulting pattern is more than just a doubling of the result of one stone's concentric ripples.

All objects have a quantum mechanical wave function. When two electrons are brought together such that their wave functions intersect, then they are described by a two-electron wave function. In the case of two rocks tossed into the pond, if the stones are identical and both are tossed into the water in the exact same way, then

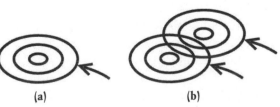

(a)   (b)

Figure 29: *Cartoon of the wave patterns observed on the surface of a pond when one rock is tossed into the water (a), and when two rocks are simultaneously tossed, near but not touching each other (b).*

the interference pattern that is observed does not depend on which rock was tossed on the left and which on the right. Similarly, in atomic physics, nothing that we can measure, such as the wavelength of light emitted from transitions between quantized energy states, can depend on any artificial labeling of the electrons. In the case of the stones in the water, they are indeed distinguishable, for we can refer to the stone on the left and the stone on the right in a meaningful way. Heisenberg tells us that it is fruitless to try to specify the location of the electron more precisely than the extent of its de Broglie wave. When two de Broglie waves overlap, concepts such as "left" and "right" become irrelevant, and all we have is the composite two-electron wave function.

Say I have two electrons, which I will creatively call electron 1 and electron 2. I bring them together so that their wave functions intersect. The electrons are indistinguishable, and no measurements can depend on which one is labeled "electron 1" and which one is "electron 2." Are there any differences at all between them at this point? Indeed yes! The two electrons, 1 and 2, have identical electrical charges and identical masses, but they can have different intrinsic angular momentum. Both electron 1 and electron 2 can have spin values of $+\hbar/2$, or both could have a spin value of $-\hbar/2$, or one could have a spin of $+\hbar/2$ while the other has spin of $-\hbar/2$. These different values of spin will be crucial for understanding solid-state physics.

Think about a ribbon, one side of which is black and the other of which is white. The ribbon represents a single electron, and if I hold the ribbon so that the white side is facing you, it indicates that the electron's spin is $+\hbar/2$, while if the black side is shown this means the spin is $-\hbar/2$. Now, if I hold two ribbons far away from each other, I can easily distinguish them—one is on the right and the other is on the left. Bring them so close that their waves overlap and I can no longer tell them apart. In this case I can describe them both with a single, longer ribbon. I can still represent the case where one electron has spin of $+\hbar/2$ and the other has spin of $-\hbar/2$, by having my right hand hold the ribbon with the white side facing out and my left hand hold the ribbon's black side facing out. Figure 30 shows a ribbon where both ends have the white side facing out, indicating that both electrons have a spin of $+\hbar/2$. The arguments presented in this figure are a modification of those made by David Finkelstein, as described in Richard Feynman's essay "The Reason for Antiparticles." I need hardly stress that the "ribbon" is simply a metaphor that will, I hope, assist in the visualization of a two-particle wave function, and is not intended as a literal representation.

The ribbon in Figure 30 represents a two-electron wavefunction with both electrons having a spin of $+\hbar/2$. The fact that the two electrons are so close that they are described by a single wave function is represented by the fact that I use one ribbon for both electrons. Any change to one electron is thus communicated to the other. What if I switch their positions, so that I move the right-hand side to the left and the left passes to the right? If I do this—without letting go of either end of the ribbon—then by switching their locations, I will add a half twist to the ribbon (Figure 30b). This is not the same situation I started with—as the ribbon has a half twist that it did not have before switching their positions. One can tell from inspection of the ribbon that a swap from left to right has occurred.

And that's it. That's the heart of Fermi-Dirac statistics, which governs the way electrons interact with one another and is the basis of the periodic table of the elements, chemistry, and solid-state physics.

How do I mathematically combine the wave functions for two electrons so that switching their order changes the situation, but

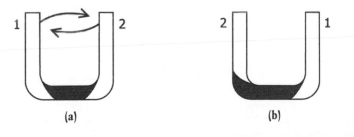

(a)   (b)

Figure 30: Cartoon sketch of a ribbon with different colors on each side, where the ribbon is presented so that each end displays the same side (the white side in this case). Switching the two ends results in a half-twist in the ribbon. Only another rotation creates a full twist in the ribbon that can now be removed by flipping one side of the ribbon twice.

making another swap restores the original state? Easy: Let the two-electron wave function Ψ be described as the difference of two functions, A and B, that is, Ψ = A – B, where A and B each depend on the one-electron wave functions at positions 1 and 2.* As in switching the two ends of our metaphoric ribbon, let's move the electron that was at one position to the location of the other electron, and conversely. In this case the wave function would be written as Ψ = B – A. The process of switching the positions of the two electrons is the same as multiplying the original two-electron wave function by (–1). If I want to get back to the original configuration, I do another switch, which brings me to Ψ = A – B again.

Nothing that I can measure should depend on which electron I label at position 1 and which one is at position 2. Now, there's no problem with having a two-electron wave function written as A – B. The fact that switching the positions is the same as multiplying the wave function by –1 will not affect any measurement we make. Remember that while the wave function Ψ contains all the information about the quantum mechanical system, it is the wave function *squared* Ψ² that gives us the probability of finding the object at some point in space and time.† It is also the wave function squared Ψ² that is used in calculating the average position (we add up all possible positions when multiplied by the probability—Ψ²—of the electron being at that position), the average momentum, and so on. And since the square of negative one is $(-1)^2 = (-1) \times (-1) = +1$, then Ψ = A – B is a physically valid way to represent the two spin $\hbar/2$ electrons. In Chapter 8, Dr. Manhattan's ability to change his size at will was ascribed to the fact that the Schrödinger equation is linear. This really becomes important here. Only for a linear equation will it be true that if A or B are separately solutions, then Ψ = A – B (or Ψ = A + B, discussed in the

---

* Those who know quantum mechanics will recognize that these two functions are $A = \Psi_m(1)\Psi_n(2)$ and $B = \Psi_n(1)\Psi_m(2)$, where $\Psi_m(1)$ is the wave function for an electron at position 1 in quantum state m and $\Psi_n(2)$ represents the second electron in a quantum state n at position 2. If the positions of the electrons 1 and 2 are interchanged, then the total wave function $\Psi = A - B = \Psi_m(1)\Psi_n(2) - \Psi_n(1)\Psi_m(2)$ acquires a minus sign. If you understood this footnote—then this book is not for you!

† As in section 2, since Ψ is a complex number, by "squaring" I mean multiplying Ψ by its complex conjugate. For our purposes this has the same effect as squaring Ψ, so we use the simpler notation of Ψ².

next chapter), will also be a valid solution of the Schrödinger equation.

Right off the bat there's a big consequence of writing the two-electron wave function as $\Psi = A - B$. What happens if I try to make both electrons be at the same location, or have both electrons in the same quantum state (when they are close enough to overlap and are described by a single two-electron wave function), so that the function A is equal to the function B? Then the two-electron wave function would be $\Psi = A - B = 0$ when $A = B$. If $\Psi = 0$, then the square of the wave function $\Psi \times \Psi = \Psi^2 = 0$ as well. Physically, this means that the probability of finding two electrons at the same place in the exact same quantum state is zero—as in, this will *never* happen. Recall in Chapter 8 our discussion of quantum mechanical tunneling. In a tunneling situation an electron in one metal, separated by the vacuum of empty space from another metal and not having sufficient energy to arc or jump from one metal to another, may nonetheless find itself in the second material. We pointed out that even though the probability for the electron to be outside of metal may be very small, as in one chance out of a trillion, there was still *some* chance of finding the electron in the second metal. The only time something will *never* be observed is if the probability of it happening is exactly zero. If something can never be observed, in physics we say that it is forbidden.

Right away, from the fact that electrons have an intrinsic angular momentum of $\hbar/2$, we can understand the structure of the periodic table of the elements. In Chapter 6 we discussed the solutions to Schrödinger's equation when the potential V is that of the electrical attraction between the negatively charged electrons and the positively charged nucleus. Schrödinger found that there were a series of possible solutions corresponding to different energy states that we argued were not unlike a series of rows of seats in a classroom, sketched in Figure 15. Some seats are close to the front of the classroom, while there are other rows farther from the front of the room. The configuration of the rows of seats depends only on the attractive force between the positive nucleus and the negative electron. We now understand why all the electrons in an atom don't just pile up in the chair in the front row, which is the lowest-energy quantum state available. For if they were to do that, then all of the electrons would be in the same location in the same quantum state, and as we have just shown, the probability of that happening is zero.

There's a fancy term used to describe the fact that no two electrons can ever be in the same position in the same quantum state—the Pauli exclusion principle. Wolfgang Pauli, one of the founding fathers of quantum mechanics, postulated this principle in 1925 in order to account for the configuration of electrons in elements. Hydrogen with one electron has the lowest energy state occupied, shown in Figure 31a. As there is only one electron in this element, it is exempt from the exclusion principle. The next element in the periodic table is helium, with two electrons. We now extend this physical analogy and propose that each "seat" in the auditorium is actually a "love seat" that can accommodate two electrons, provided that they face away from each other (that is, as long as one is spin "up" and the other is spin "down."* As in Figure 31b, both of these electrons can reside in the lowest energy state, as long as one has a spin value of $+\hbar/2$ and the other has a spin of $-\hbar/2$, since each spin state counts as a different quantum state. As there are no other possible spin values, a third electron in lithium (the next element up the table, shown in Figure 31c) will have to reside in the next higher energy state. If all three electrons were to reside in the lowest energy state in lithium, then there would be at least two electrons both with spin = $+\hbar/2$ or spin = $-\hbar/2$, and the probability of this occurring is $\Psi^2 = 0$. Carbon, shown in Figure 31d, has six electrons—two sit in the ground state, and the remaining four sit in the next highest "row of seats"—and is able to form chemical bonds in a wide variety of ways. By forming these bonds, the carbon atom and the other atoms it chemically interacts with lower their energies, compared to their unbonded states. If all of carbon's six electrons could drop down into the lowest energy state, there would be no energetic advantage to forming chemical bonds with other atoms. Consequently, there would be no methane, no diamond, no DNA, without the Pauli exclusion principle.

---

* For simplicity, the nucleus in Figure 31 is represented by a single positive charge, while in fact there are multiple protons in all nuclei except hydrogen, from two in helium (Figure 31b) to thirteen in aluminum (Figure 31e). I have also not attempted to represent the changes in the spacing of the rows for the different elements.

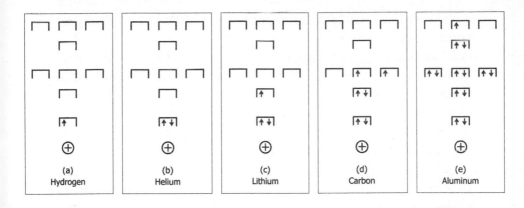

Figure 31: *Representation of the allowed quantum state solutions to the Schrö-edinger equation for an electron in an atom as a set of seats in a classroom. The Pauli principle indicates that each seat can accommodate two electrons provided they have opposite spins. Shown from left to right are the occupied quantum states for atoms containing one, two, three, six, and thirteen electrons, corre-sponding to hydrogen, helium, lithium, carbon, and aluminum, respectively.*

Consider an atom such as aluminum, with thirteen electrons, shown in Figure 31e. All but one of these electrons are arranged in $+\hbar/2$ and $-\hbar/2$ pairs, and thus only this last electron can participate in chemical bonds. The other twelve electrons are chemically inert and form the inner core of the aluminum atoms. Not that we can't make use of these inner electrons. The Pauli principle forces the electrons to reside in higher and higher energy states, equivalent to having some students sit in rows far from the front of the class-room, even when the atom is in its lowest energy configuration. If we could knock one of the electrons out of the ground state (a row close to the front of the lecture hall), then an empty position sud-denly would open up, as if we had ejected a student sitting in a front-row seat. A student sitting in one of the upper rows could then jump down into the newly vacant seat. Just such a situation can arise when a high-energy beam of electrons strikes an atom. In that case, when one of the outer electrons falls down to occupy the lower energy state, it can emit an X-ray photon during the transi-tion. This is in fact a very efficient way to generate X-rays, and most dental X-ray machines employ electron currents striking a copper target to create the penetrating radiation.

How do the last few electrons that are not residing in paired quantum states, and are thus available to participate in chemical bonds, combine with those from neighboring aluminum atoms to hold all trillion trillion atoms together in a solid piece of aluminum? How do the last unpaired electrons between carbon atoms in diamond combine to bind this rigid insulator? In both cases, the electrons arrange themselves to satisfy the Pauli exclusion principle, though the resulting material properties in aluminum and diamond could not be more different.

One easy way to satisfy the Pauli principle is to never let the electrons be at the same place at the same time. If I have a line of atoms, and next to each atom is a barrier, then I can place an electron inside each theoretical box (we'll see soon what this "box" really is), and all these negative charges can be in the same quantum state. This does not cause any problem, for by creating a series of containers for each electron, I have in principle made them distinguishable. I can tell which electron is in the box on the right and which on the left, just as I could tell apart the two stones that I tossed into the pond. The walls of the boxes prevent the de Broglie waves of each electron from overlapping with those of its neighbors, so the trillion trillion electrons can all be in the same quantum state, as the total wave function is just the one electron function repeated a trillion trillion times. Each electron is described by its own ribbon, as shown in Figure 30, and no ribbon is used for more than one electron. When I calculate the average energy of each electron in a box, assuming the width of the box is the spacing between atoms in my solid, I arrive at a number of about three electron Volts (the exact value obviously depends on all sorts of details of how the atoms in the solid are arranged—termed the crystalline configuration). This is the energy I would have to give to an electron to remove it from the box. Of course, as each electron has two possible values of spin, I can actually put two electrons in each box (each box contains a love seat), as long as they have intrinsic angular momentum $+\hbar/2$ and $-\hbar/2$.

Consider carbon, shown in Figure 31d. Carbon can easily "rearrange the seats in the rows" of the four upper electrons, mixing the quantum mechanical wave functions to form differing configurations of quantum states that allow for a variety of chemical interactions. Carbon can form strong bonds in a straight line, in proteins and DNA; it can form graphite, with three strong bonds in a plane and one weak bond above or below the plane, which is why graphite can be easily peeled apart when used in a pencil, for example; and when the "seats" are configured to form four equally strong bonds, we call this form of carbon "diamond." In each case, the carbon atom has four electrons that are capable of participating in chemical bonds, represented by four boxes, each of which holds one electron. The Pauli principle tells us that each box can hold a second electron, provided it has an opposite spin from the first. When two carbon atoms come close enough to each other that the quantum states containing these unpaired electrons overlap, the two electrons can be represented by a two-electron wave function. It turns out that each of the unpaired electrons can lower its energy if the electrons fill up the love seats in each box (Figure 32a). That is, I must add energy to the atoms to remove the electrons from these boxes, and restore each one to its unpaired state. The overlapping electron wave functions form a chemical bond between the atoms, holding them together in the crystalline solid. The Pauli principle is satisfied by the localization of the electrons in space. In a diamond crystal, each carbon is surrounded by four other carbon atoms in a tetrahedral arrangement, and their unpaired electrons can occupy the second seat in the first atom's boxes (Figure 32b).

But there is another way to satisfy the Pauli exclusion principle. Let's say that there are no boxes, and I let the electrons wander over the entire solid. In this case two electrons can be at the same place at the same time, so I have to ensure that they are each in different quantum states. How many different lowest-energy quantum levels are available for the last unpaired electron, such as the thirteenth electron for aluminum, shown in Figure 31e? As many as there are atoms in the solid. Since I have given up having any knowledge of where the electrons may be, I can compensate by having the electrons reside in states that have a well-defined momentum. I can have a matter wave with a very large wavelength, equal to the entire length of the solid. Since (momentum) × (wavelength) = Planck's constant, this large wavelength corresponds to a very small momentum, and hence energy. At the other extreme, the smallest wavelength that can be constructed corresponds to the distance between atoms in the solid. This is very short, so the momentum

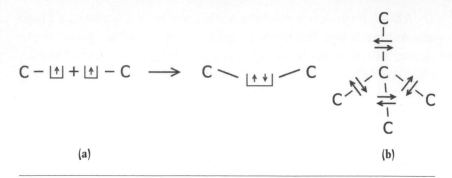

(a)   (b)

Figure 32: *Sketch of the lowering in energy when two unpaired electrons from adjacent carbon atoms overlap and form a carbon-carbon bond (a). Also shown is a sketch of the configuration of carbon when in the diamond configuration, allowing four chemical binds with its neighbors, in a tetrahedral orientation (b).*

of this matter-wave is high. The highest energy of this shortest wavelength is *also* about three electron Volts, again, depending on the details of the atomic configurations in the solid.

So, whether the electrons are put inside boxes in each atom, or allowed to roam over the solid, we still wind up with an energy of about three electron Volts. However, in the second situation, where the electrons can move around the solid in discrete momentum states, three electron Volts is approximately the energy of the *most* energetic electron, while when the electrons are placed in boxes, three electron Volts is roughly the energy of *each* electron. The average energy of an electron in the free-to-roam case is less than three electron Volts, and in fact will be closer to 1.5 electron Volts (recall the class from our discussion of the Heisenberg uncertainty principle, where every student had a different exam score, from 0 to 100. In this case the average grade was 50 percent). For the electrons-in-a-box situation, every electron has the same energy, so the average energy is also three electron Volts (if every student scores a perfect 100 percent, then the class average is also 100 percent). Consequently, depending on its chemical composition, the solid as a whole can lower its energy by letting the electrons wander around the crystal. This won't be true for all solids. Some materials will be able to lower their total energy by keeping every electron localized in boxes around each atom. We call the free-to-roam cases "metals," and the electrons-in-a-box materials "insulators."

And that's how quantum mechanics explains solid-state physics. At very low temperatures, all solids either conduct electricity or they don't. We call the first case metals, and the second are insulators (the distinction between insulators and semiconductors is most relevant around room temperature, and I defer for now a discussion of the differences between the two).

Metals such as aluminum are good conductors of electricity because the outermost electrons satisfy the Pauli principle by residing in momentum states and are free to move around the entire solid, while insulators keep theirs in boxes (bonds) around each atom. To remove a metal atom from the solid, I must first grab one of the free-range electrons and localize it on a positively charged atom—in essence, put it in a box so that I can pull the neutral atom out of the solid. But this costs me a few electron Volts of energy, and this can be considered the binding energy holding the atoms together in the metal. There are no directional bonds between atoms, so it is easy to move atoms past each other, which is why metals are easy to pull into wires or pound into thin sheets, without losing their structural coherence. If light is absorbed by the solid, there is always a free electron that can absorb its energy and reemit it back again, which is why metals are reflective and shiny. The sea of free electrons makes metals good conductors of both electrical current and heat.

Insulators, on the other hand, such as diamond, have all the electrons tied into bonds between the atoms (the two electrons per box as discussed earlier). They are thus poor conductors of electricity and can conduct heat only by atomic vibrations (sound waves). The electrons in the box can assume only specific energy states, like electrons in atoms. Consequently, if you shine light on an insulator that does not correspond to an allowed transition, it will ignore the photons. This is why some insulators, such as diamond or window glass, are transparent to visible light. The details of the materials' properties are very sensitive to the configurations of the boxes in which the electrons reside, which is why changes in crystal structure—say, when carbon transforms from graphite to diamond—can yield big variations in optical and electrical properties. The boxes here are the directional, rigid chemical bonds between the atoms (Figure 32b), and changes in the type of these bonds and the chemical constituents lead to big variations in crystal structure and rigidity (diamond is very stiff, while graphite is so soft, we use it for pencil lead). All the differences between insulators and metals can be understood at the most basic level by whether the last few unpaired electrons of the atoms in the solid satisfy the Pauli exclusion principle by localizing themselves in real space (insulators) or in momentum space (metals).

Thus, from playing with a ribbon, with one side black and the other side white, we see why the world is the way it is. Note that not everything has an intrinsic angular momentum equal to $\hbar/2$. Some objects, such as helium atoms or photons, have spin values of either 0 or $\hbar$. This seemingly small difference leads to superconductors and lasers.

CHAPTER THIRTEEN

All for One and One for All

Bert Hölldobler and **Edward O. Wilson,** in *The Superorganism—The Beauty, Elegance, and Strangeness of Insect Societies*, propose that colonies of wasps, ants, bees, or termites can be considered as a single animal. They argue that each insect is a "cell" in the "superorganism": foragers are the eyes and sense organs, the colony defenders act as the immune system, and the queen serves as the colony's genitalia.* An important difference between a superorganism and a regular animal is that the colony lacks a centralized brain or nervous system. Rather, each colony has its own rules for local interactions among the insects that govern its organization and size. In this way the colony can achieve levels of development that are far beyond the capabilities of the individual insects were they to act alone. As readers of science fiction pulps know, this is the mechanism (or at least one of them) by which humans defeat an alien invasion.

In Theodore Sturgeon's 1958 science fiction novel *The Cosmic Rape* (an abridged version was simultaneously published in *Galaxy* magazine with the title *To Marry Medusa*), an alien intelligence applies an unconventional approach to its efforts to conquer Earth. (Spoiler alert!) The alien, in fact a cosmic spore, is capable of con-

---

* This notion that organisms with specialized skills could serve as analogs of sensory organs or other bodily or psychological functions in a gestalt organism was anticipated in Theodore Sturgeon's 1953 science fiction novel *More Than Human*.

trolling the intelligence of a single human. It is surprised to discover that the people of Earth are not in mental contact with one another, given that the planet is covered with complex structures such as buildings, bridges, and roads. In the spore's experiences conquering other planets that contained advanced infrastructure, such architecture is possible only when the primitive intelligences of the individual agents interact in a cooperative manner, as in a colony of ants or a hive of honeybees. The invading spore had never encountered a species for which a single agent is capable of designing a bridge or building on is or her own, and it thus assumes that a previously existing collective connection has been severed. The spore sets upon a plan to reestablish this connection and force all humans to think and work together in unison. Unfortunately for the alien entity, it succeeds in its plan. Once it finds itself dealing not with the intelligence of a single human, but with the collective consciousness of all several billion humans, the "hive mind" of humanity quickly devises an effective counterattack, destroying the alien spore. In this way Sturgeon has described the cooperative behavior of a Bose-Einstein condensate.

In Sturgeon's novel, the alien spore creates from humanity a macroscopic quantum state, with a single wave function containing all the information about its constituent elements. Any change in one element, in Sturgeon's case a human being, is instantly transmitted to every other element in the wave function, that is, the rest of humanity. Such situations occur frequently in the real world through quantum interactions between pairs of electrons in a superconductor or helium atoms in a superfluid. These collective states involve particles whose intrinsic angular momenta are multiples of $\hbar$, rather than $\hbar/2$. Particles with intrinsic angular momenta that are whole-number multiples of $\hbar$ are called bosons, as they obey a form of quantum statistics elaborated by Satyendra Bose and Albert Einstein, termed Bose-Einstein statistics.

In the previous chapter we discussed fermions, for which the angular momentum could be either $+\hbar/2$ or $-\hbar/2$, but not any other value. This is intrinsically asymmetric, as we can distinguish a top twirling clockwise from one rotating counterclockwise. We represented this situation, when the two fermions are so close that their wave functions overlap, with a ribbon with one side black and the other white. The significance of the two colors was that we could readily distinguish the spin = $+\hbar/2$ electron from the spin = $-\hbar/2$ electron, as we could the black and white sides of the ribbon. But experiments have revealed situations where the intrinsic angular momentum can have values of 0 or $\hbar$ or $2\hbar$, and so on, but not any fractional values. Let's consider the case of quantum objects with a spin of zero first, and then turn to spin = $\hbar$ particles such as photons.

As the fundamental building blocks of atoms—electrons, protons, and neutrons—are all fermions, what sort of object would have spin of zero? One example is a helium atom. A helium nucleus has two protons and two neutrons, each having spin = $+\hbar/2$ or $-\hbar/2$. As the two protons are identical, in their lowest energy state in the nucleus they would pair up, $+\hbar/2$ and $-\hbar/2$, for a total spin of zero, as would the two identical neutrons. Similarly, the two electrons are spin paired, as indicated in the sketch in Figure 31b. Consequently, the total intrinsic angular momentum of a helium atom, when in its lowest energy configuration, has a spin value of zero.

Particles with zero value of spin are symmetric, in that we cannot describe the rotations as clockwise or counterclockwise. When two such particles are brought so close that their wave functions overlap, we will represent them by a ribbon whose sides are both white. I once again stress that the ribbon is employed as a metaphor for the resulting two-particle wave function, and as such, certain issues are being ignored here that would only distract from our discussion.

Let's repeat the experiment with the ribbon from Chapter 12, only now using a ribbon with both sides the same color, for example, white (Figure 33). I can hold each end, and obviously a white side of the ribbon faces out (Figure 33a). Now, without letting go of either end, I will switch their positions, as before. The end of the ribbon on the left is now on the right, and vice versa. This procedure has introduced a half twist in the ribbon, as in Chapter 12 (Figure 33b). Of course, now both sides are still white. I can undo the half twist by flipping one end of the ribbon around, so that the back side turns outward (Figure 33c). When the ribbon had one side white and the other side black, this was a forbidden operation, as it changed the state of the ribbon (where before both ends had

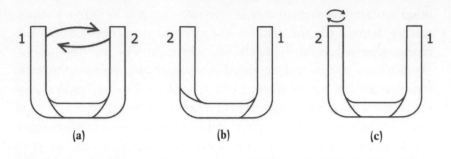

Figure 33: *Cartoon sketch of a ribbon with the same color on each side (a). Switching the two ends results in a half-twist in the ribbon (b) that can be undone by rotating one side of the ribbon (c), restoring the original configuration.*

white facing out, one side would then have had a black side facing out). But if both sides of the ribbon are white, then this symmetry means I can flip one end of the ribbon and I have not changed anything except undoing the half twist. The important point is that a white ribbon can be restored to its original state following a single rotation, while the black/white ribbon requires two rotations to bring the original configuration back.

This symmetry indicates that the two-particle wave function for spin = 0 particles, such as helium atoms, as well as spin = $\hbar$ photons, termed bosons, can be written as the sum of the two functions A and B, $\Psi = A + B$, rather than $\Psi = A - B$, for fermions.* As before, A and B depend on the product of the one-particle wave functions at positions 1 and 2. Now the two-particle wave function $\Psi = A + B$ is unchanged if the positions of particles 1 and 2 are switched, in which case Ψ would be given by $\Psi = B + A$. But this is just the same as $\Psi = A + B = B + A$. When two particles for which the intrinsic angular momentum has values of spin = 0 or spin = $\hbar^\dagger$ are brought close enough to each other that their de

---

* As before in Chapter 12, the functions A and B are products of the one-particle wave functions, with $A = \Psi_m(1)\Psi_n(2)$ and $B = \Psi_n(1)\Psi_m(2)$, where $\Psi_m(1)$ is the wave function for a boson at position 1 in quantum state m and $\Psi_n(2)$ represents the second identical boson in a quantum state n at position 2. If the positions of objects 1 and 2 are interchanged, then the total wave function $\Psi = A + B = \Psi_m(1)\Psi_n(2) + \Psi_n(1)\Psi_m(2)$ is unchanged.

† While the illustration with the ribbon does not apply for spin = $\hbar$ (or $2\hbar$, and so on) particles, the mathematical arguments above about symmetric two-particle wave functions for these integral spin particles is identical to the spin = 0 case.

Broglie waves overlap, the resulting two-particle wave function is just the sum of the functions A and B, which are in turn functions of the one-particle wave functions.

What is the consequence of writing the two-electron wave function as $\Psi = A + B$? Recall that for fermions such as electrons, the fact that the two-electron wave function is $\Psi = A - B$ meant that the probability is exactly zero that both electrons would be in the same quantum state, for which $A = B$. For bosons, $\Psi = A + B$ indicates that the probability is large exactly when both particles are in the same quantum state, when $A = B$. Because when $A = B$, then $\Psi = A + A = 2A$ and the probability density $\Psi^2 = (2A) \times (2A) = 4A^2$. For a single particle in state $\Psi = A$ the probability density would be $\Psi^2 = A \times A = A^2$. For two single particles the probability would be $A^2 + A^2 = 2A^2$. So just by bringing a second identical particle near the first, the probability that they would both be found in state A is double what it would be for the two particles separately. While the probability is not 100 percent that they will both be in the same state, it is enhanced compared to the single-particle situation. A larger probability of both particles being at the same location in the same quantum state indicates that it is more likely to occur than not.

As the temperature of a system is reduced, the particles will settle down into lower energy states. If we had particles that were somehow distinguishable, for example, if their wave functions did not overlap so we did not have to worry about Fermi-Dirac or Bose-Einstein statistics, then at low temperatures we would find many particles in the lowest energy state, some in the next available quantum level, a few more in the next higher level, and negligible occupation of very high-energy states. For fermions, such as electrons in a solid, only two electrons can occupy the lowest energy level (one with spin $= +\hbar/2$ and the other with spin $= -\hbar/2$), regardless of temperature. In contrast, bosons will have an enhanced probability of collecting into the lowest-energy ground state at low temperatures, relative to the distinguishable particle case. For these particles, the rule of one particle per spin orientation per seat (valid for fermions) is thrown out, and one can have many particles dog piling into a single state. These spin $= 0$ or spin $= \hbar$ particles obey statistics described by Bose and Einstein, and this settling into the ground state is termed Bose-Einstein condensation.

Why do we need to go to low temperatures to see this condensation? If the particles are very far apart, then there will be little or no overlap of their wave functions, and the whole issue of indistinguishable particles is irrelevant. Temperature is just a bookkeeping device to keep track of the average energy per particle, so the lower the temperature, the less kinetic energy and the lower the momentum. From de Broglie's relationship, a low momentum corresponds to a long matter-wavelength. If the particles involved have long de Broglie wavelengths, it will increase the opportunity for the waves of different identical particles to overlap. Similarly, confining the particles to a small volume also increases the possibility for interactions among wave functions. Consequently, low temperatures and small volumes (achieved by squeezing the system at high pressures) help induce Bose-Einstein condensation.

What are the special attributes of a Bose-Einstein condensate? We have considered the case of two identical bosons whose wave functions overlap such that they can be described by a single, two-particle wave function. As the temperature of a gas of bosons is lowered, millions of identical atoms' wave functions overlap, all in the same quantum state. We thus obtain one single wave function that describes the behavior of millions of atoms. In this way the individual indistinguishable bosons behave as a single entity, and whatever happens to one atom is experienced by many. The Bose condensate is not unlike the demonically possessed children in the 1960 science fiction film *Village of the Damned*. The fair-skinned, blond children play the role of indistinguishable particles, and the fact that knowledge gained by one child is instantly shared with all is a natural consequence of the multiparticle wave function that describes this collective phenomenon.

True condensation, confirming the theoretical predictions of Bose and Einstein from the 1920s, was experimentally observed by Eric Cornell and Carl Wieman in 1995, and independently by Wolfgang Ketterle the same year, a feat for which they shared the 2001 Nobel Prize in Physics. Their investigations involved thousands of particular isotopes of rubidium or sodium, cooled to temperatures below a millionth of a degree above absolute zero. While these

Bose-Einstein condensates are ephemeral quantum objects, difficult to obtain and to probe, there are more robust systems that owe
their striking properties to the clustering of bosons into a single
low-energy quantum state.

As pointed out earlier, helium is an example of an atom that is
characterized by total spin of zero, and is thus a boson. The two
electrons in helium are spin paired in the ground state (Figure 31b),
and helium thus does not have strong chemical interactions with
other atoms—a feature it shares with other elements whose electrons are paired up in completely filled "rows," such as neon and
argon, the inert, or noble, gases. These elements consequently remain gases until their temperature is so low that small fluctuations in their electrical charge distribution induce weak electrical
attractions. Helium interacts so weakly with other helium atoms
that it does not form a liquid until 4.2 degrees above absolute zero.
If cooled even further at normal pressures, it does not form a solid
but rather undergoes a quantum transition, where some of the
atoms condense into the ground state.

Suppose the temperature of liquid helium is lowered all the
way to absolute zero. We would expect that the helium would
eventually become a solid, but it in fact remains a liquid, thanks
to the uncertainty principle. At low temperatures, when the wave
functions overlap, the uncertainty in the position of each atom is
low. There is thus a large uncertainty in the momentum of each
atom, which contributes to the ground-state energy of the helium atoms (called the "zero-point energy"). The lower the mass of
the atom, the larger this zero-point energy, and for helium this
contribution turns out to be just big enough to prevent the atoms
from forming a crystalline solid, even at absolute zero. Hydrogen
has an even lower mass than helium, but it forms a solid at 14
degrees above absolute zero due to strong electrical interactions
between hydrogen molecules, while for heavier elements the uncertainty in the momentum of each atom is not enough to over-
whelm the tendency to form a solid at low temperatures. While
helium does not form a solid at normal pressures (if you squeeze
the liquid, you can force it to form a crystal), it does undergo a
"phase transition" at 2.17 degrees above absolute zero, as some of
the helium forms a condensate in the ground state.

What would be the properties of a fluid for which some of the atoms have condensed into a single quantum state? One surprising feature would be that the fluid would have no viscosity! Viscosity describes the internal friction all normal fluids have; you can think of it as resistance to flow. Water has a pretty low viscosity, and molasses and motor oil have much larger viscosities. A fluid with no viscosity would, once it started moving, continue to flow at a constant speed through a hose without continued applied pressure. Such a state is termed a superfluid, for it does what a normal fluid does—but with the power of quantum mechanics!*

Experimentalists in 1965 rotated a spherical container of liquid helium at 4 degrees above absolute zero about an axis passing through its center. The sphere was packed with glass particles, so the fluid would have to move through the small pores and gaps between the beads. The liquid helium, not yet a superfluid at this higher temperature, began to swirl along with the container. The temperature of the helium was then lowered to below 2.17 degrees above absolute zero, at which point some of the helium condensed into the superfluid state. When the container's rotation was then stopped, the superfluid continued to move with no change in speed. When you stop stirring your coffee, the fluid comes to rest within a few seconds, but the superfluid helium maintained its circulating motion for hours, until the researchers eventually stopped the experiment.

If this system were warmed higher than a temperature of 2.17 degrees above absolute zero, then the superfluid would transform into a normal fluid, and it would rapidly cease rotation. The low temperature is crucial. At low enough temperatures, the lower momentum of the helium atom corresponds to a long de Broglie wavelength. There is then sufficient overlap among the many helium atoms' wave functions that all of the atoms can be described by a single macroscopic quantum mechanical wave function. To slow down even one helium atom in the condensate, it is necessary to

---

* Strictly speaking, the transition that helium makes as it forms a superfluid, or when the electrons form Cooper pairs in a superconductor, is not technically an example of Bose-Einstein condensation. The distinctions between a true Bose-Einstein condensate and the superfluid or superconducting state are technical, and for our purposes we may take them to be the same.

decelerate the entire multi-atom wave function, and provided the rotation is not too fast, there isn't sufficient energy to do this. So the superfluid keeps on rolling.

There is an electrical analog to superfluidity that is found in many metals and even some nonmetallic materials, termed "superconductivity." A wire that is a superconductor has no electrical resistance and is analogous to a garden hose through which a superfluid flows. Any nonviscous fluid pushed into one end of the hose would exit at the other end with the same velocity, no matter how long or clogged the hose, even if the tubing circled the equator. While we don't know whether the electrical current circulating in a superconducting loop will flow forever, experiments have confirmed that even after a year, the supercurrent in a closed ring has decreased from its initial value by less than one part in one hundred trillion.

In a normal electrical conductor, an externally applied voltage induces an electrical current. The smaller the resistance of the conductor, the greater the current for a given voltage. In the standard water-flow analogy for electrical circuits, water pressure plays the role of voltage. The greater the pressure, the more of a push exerted on the water. The flow of water out of the faucet through a garden hose is analogous to the electrical current. If the hose has imperfections and bumps along its length that make it difficult for the water to flow, this would be analogous to the electrical resistance of the wire. The water loses energy through collisions with these partial blockages, as well as with the walls of the hose, so a constant pressure is needed to maintain a steady water flow out of the end of the hose. Similarly, as the electrical current collides with imperfections in the wire, some of the current's energy is lost. This is why a constant push (a voltage) produces a constant flow (electrical current) rather than an accelerating flow (Newton's second law, that force = mass × acceleration, would suggest that if the force is constant, then the acceleration, that is, the rate of change of the electron's speed, should also be constant). Superconductors have no electrical resistance, so that a current, once started, will continue unchanged without an applied voltage, just as the helium atoms in a superfluid are able to translate without viscosity.

Currents in metals are carried by electrons, not helium atoms. Remember that electrons are fermions that have intrinsic angular momentum of $\hbar/2$. In order to observe Bose-Einstein condensation, the electrons must form a composite particle consisting of two electrons, one having spin $= +\hbar/2$ and the other with spin $= -\hbar/2$. Thus, the two-electron composite would have a net intrinsic angular momentum of zero and would therefore be a boson. As such, at a low enough temperature, these paired electrons would condense into a ground state and be able to flow without resistance.

Electrons are negatively charged, and as two negative charges repel each other, the question is, Why would two electrons bind together to form a composite particle that acts as a boson? The answer lies in the positively charged atoms, termed "ions," that make up the metal. Recall from the preceding chapter that in a metal such as lead, the last unpaired electrons from each atom reside in quantized momentum states. The electrons are free to roam over the solid but can do so in well-defined energy states. As the metal atoms were initially electrically neutral, if an electron leaves the immediate vicinity of its atom, it leaves behind a positively charged ion (an "ion" is an atom with a net electrical charge due to the removal or addition of electrons). These metallic ions form ordered arrays and comprise the crystal. As a negatively charged electron moves through the metal, the positively charged ions are attracted to it. The positive ion is too large to leave its position in the crystal, but it strains toward the negatively charged electron, slave to the electrostatic attraction between them.

As the electron speeds along, it leaves in its wake a trail of positively charged ions that are pulled along its trajectory, not unlike the way metallic objects bend towards Magneto (the mutant master of magnetism) from the X-Men comic books when he employs his mutant power. In time the ions would be repelled from each other and return to their normal crystalline locations. At temperatures less than 7 degrees above absolute zero, the lead ions move slowly, and this positively charged channel in the wake of the first electron can persist long enough for a second electron to be attracted into this positive valley. That is, the first electron polarizes the positive ions in the lattice, and a second electron is attracted to this positively charged channel and follows the same path. In this way the two negatively charged electrons are bound together and form what is known as a Cooper pair (after Leon Cooper, who first theoretically showed that such a binding mechanism could operate in metals at low temperatures). The lowest energy configuration corresponds to two electrons with spins of $+\hbar/2$ and $-\hbar/2$, respectively, so the Cooper pair formed from the two bound electrons has an intrinsic angular momentum of zero and acts as a boson.

Once at least some of the electrons in a metal start acting like bosons and condense into a low energy state, superconductivity is observed. When the wave functions for the many Cooper pairs overlap, they form a single multiparticle wave function. In a normal metal, collisions with vibrating atoms or defects in the metal cause the current to lose energy, which is why a constant voltage is needed to maintain a uniform current. In order to slow down a supercurrent consisting of a condensate of overlapping Cooper-paired electrons, the collisions must break apart a Cooper pair, also changing the energies of all the overlapping pairs, and at low temperatures and moderate currents this is not energetically possible. The Cooper-paired electrons are able to carry electrical current (provided it isn't too high) without any resistive loss, just as the helium atoms in a superfluid are able to flow (but not too fast) without viscosity.

There are many free-range electrons in a metal, but not all of them have to form Cooper pairs for the metal to exhibit superconductivity. What about the other electrons that may not bind up in pairs to form boson composite particles? They still have a normal resistance, but their contribution is shorted out by the supercurrent. If I have two roads to a destination, one that is a bumpy, unpaved dirt road with a speed limit of 5 miles per hour, and another a sleek superhighway with no upper speed limit, I will take the second road.* Any electrical current in a metal, which has been cooled below the temperature at which Cooper pairs form, will be carried by the superconducting paired electrons. Similarly, not all the helium atoms in a superfluid reside in the ground-state condensate. As long as some of the particles in the superconductor or superfluid are in a lower energy condensate, they will exhibit cooperative behavior.

---

* And that has made all the difference!

Superconductors do not just carry electrical current with no resistance whatsoever—they also are perfect diamagnets. This means that they resist any externally applied magnetic field. Some metals are attracted to magnets, while others are actually repelled. Gold and silver are examples of this latter type of metal. If you are able to pick up your "gold" jewelry with a refrigerator magnet, you should probably look into a refund (or at least check to see whether the jewelry is filled with chocolate). The internal magnetic fields of the gold atoms polarize in the opposite direction to an external magnetic field, such that they develop a north pole that faces the applied north pole. As north repels north, the gold ignores the magnet, or if the applied magnetic field is strong enough, the gold is pushed away from the outside magnet through its diamagnetism.

Superconductors are *perfect* diamagnets, as they can set up electrical currents that generate magnetic fields that *exactly* cancel out inside the solid all of the externally applied magnetic field. As these materials have no electrical resistance, once the current is started, it can continue indefinitely as long as the outside magnetic field is applied—which would make superconductors ideal materials from which to construct rails for magnetically levitating trains. The drawback, at present, is the ultralow temperatures necessary to induce superconductivity in most metals. In Section 6 I discuss materials termed "high-temperature superconductors" that show superconductivity at much higher temperatures (though not yet at room temperature) and turn out to not even be metals.

Aside from certain elementary particles generated at particle accelerators or in cosmic-ray showers, most bosons that have mass are composite particles such as the helium nucleus or Cooper pairs of electrons. There is, however, a very common mass*less* particle that has an intrinsic angular momentum = $\hbar$ and obeys Bose-Einstein statistics—light!*

Recall our discussion in Chapter 1 of Max Planck and how his explanation of the spectrum of light emitted from hot objects, shown in Figure 2, ushered in the quantum age. Measurements of the intensity of light given off from an object, as a function of the

---

* When the intrinsic angular momentum is measured relative to the direction of the photon's motion

frequency of the light, found that very little light is emitted at low and high frequencies and that the light intensity peaks at a frequency that depends only on the temperature of the object (Figure 2 showed the light spectrum for an object with you right now). Theoretical physics calculations prior to Planck indicated that the intensity should indeed be small at low frequencies but would grow without limit as the frequency increased. We now know enough quantum statistics to see what these theorists got wrong.

A box of molecules, such as a gas, will have some total amount of energy that is indicated by the gas's temperature. A central principle of nonquantum thermodynamics is that as the gas molecules collide with one another, they share and transfer their energy, so that in equilibrium we would find that on average each molecule has an equal portion of the total energy of the gas. Of course there will be random fluctuations, so that one may see a molecule with a little more or a little less energy than the per-molecule average, but subsequent collisions with other molecules would tend to bring this molecule's energy back toward the per-molecule average. When you add up the average energy per molecule for all trillion trillion molecules in the box, you get the total energy of the gas. This not only is perfectly reasonable, but is in fact what is observed in real gases (when the quantum nature of the molecules can be ignored, that is, at high temperatures and low pressures, so that the molecule's de Broglie wavelengths do not overlap).

What if the box were filled with light, treated as extended electromagnetic waves? The atoms in the walls of the box are at some temperature and will jitter back and forth around their normal crystalline positions. It was known before quantum mechanics that oscillating electrons emit electromagnetic waves, which is the basic principle underlying radio and television broadcasting. If there is some dust in the box that absorbs and reemits light, serving the same role as the collisions between gas molecules described earlier, then each electromagnetic wave will have the same average energy per wave.

Nonquantum thermodynamics, which is the only kind that existed before Max Planck published his paper in 1900, would say that the average energy per wave is a constant multiplied by the temperature of the system. This analysis works very well for a box filled with gas molecules. For the case of the gas molecules, we add up the average energy per molecule for the trillion trillion molecules and find the total energy of the gas. A trillion trillion is a big number, but it's just a number. However, there is no upper limit on the frequencies of waves that could possibly reside in a box filled with light. A clamped guitar string when plucked has a lowest possible frequency, but there is, in principle, no upper limit on the highest frequency that can be excited. If each possible wave has the same average energy per wave, and there are an infinite number of possible waves, then the total energy of the light in the box is infinite! Fortunately, this does not happen in real objects, or else all matter would emit an infinite amount of energy in the form of X-rays and gamma rays. This would be catastrophic, which is why physicists called it the ultraviolet catastrophe.

To resolve this contradiction between calculated and observed light-intensity spectra, Planck assumed that the atoms in the walls of the box can lose energy only in steps proportional to the frequency of the light, from which the relationship Energy = $h \times$ (Frequency) was proposed. At this stage we know more about quantum mechanics than Planck did in 1900, so we can use a simpler argument than his original one to understand the observed spectrum of light emitted by all glowing objects.

The box containing light can be considered a gas of photons, each of which has an intrinsic angular momentum of $\hbar$. These photons are thus bosons and will obey the same Bose-Einstein statistics that we invoked for helium atoms and Cooper-paired electrons. For a gas of bosons, there is an enhanced probability of finding the particles in lower energy states. Most bosons will be in the lowest energy state, some will be in the next higher level, a few will be in the next higher energy state, and states very high in energy will have an exponentially small chance of being occupied.

The energy of the photon gas as a function of the frequency of light is the energy of the photon ($E = h \times f$) multiplied by the average number of photons with this energy. Most of the photons are in the low-energy states that carry very little energy. Higher-energy photon states are exponentially less likely to be occupied, so the contribution to the average energy from these photons will also be low. The resulting product of an increasing energy for a photon with a decreasing number of photons with that particular energy yields an average energy per frequency that is very low at low frequencies, peaks at some intermediate frequency, and is again very small at higher frequencies, exactly as observed.

This is akin to the payouts for a Powerball or Lotto lottery system. There the ticket holder must match all six randomly selected numbers in order to claim the grand prize jackpot. But even if no one matches all six numbers, smaller awards are possible. Those matching only three of the selected numbers will win a smaller prize, say ten dollars. Those with four matching numbers might win ten thousand dollars, and five matches would garner one hundred thousand dollars. The payout amount starts off small—many people may match one or two numbers, but they do not win any money; some will match three numbers, but the amount they win is low—fewer will match four numbers, but they have a larger payout, and very few will have selected five of the winning numbers, so there the total payout will also be lower (a large prize but with few winners). A graph of the amount paid out by the lottery agency against the size of the prize would start off small, reach some peak value, and then drop back down.

The number of gas molecules in a box is fixed when we set up the container, but the number of photons can vary, depending on the temperature. Hot objects emit very bright light (that is, give off a large number of photons), while cooler objects emit a lower number of photons. At higher temperatures the exponential tail in the number of photons extends to higher energies. The peak in the spectrum of light energy emitted by a glowing object as a function of frequency will thus depend on the object's temperature. Cold objects will have their peak at lower frequencies, and the hotter the object, the higher the frequency at which the curve peaks.

Measurements of the light spectrum of objects that can be considered blackbodies therefore provide a way to determine the temperature of very hot objects, such as the interior of a blast furnace or the surface temperature of the sun. But this technique works for cold objects as well. Space is infused with microwave radiation that is the remnant energy from the big bang creation of the universe. Measurements of the spectrum of this radiation as a function of frequency find that it beautifully fits the Planck expression if the characteristic temperature of the universe is 2.7 degrees above absolute zero. Figure 2 from Chapter 2 in fact shows the measured blackbody spectrum of the cosmic microwave background radiation, present at every point in the universe, even where you, Fearless Reader, are right now! From the measured expansion rate of the universe, we can determine that it took approximately fifteen billion years for the universe to grow and cool to its presently measured temperature. In Section 3, I showed how quantum mechanics, developed to account for the manner by which atoms interact with light, enables, through radioactive isotope decays, a determination of the age of the Earth. Now we see that quantum physics also provides an age for the oldest thing in the universe—the universe itself!

MODERN MECHANICS AND INVENTIONS

CHAPTER FOURTEEN

Quantum Invisible "Ink"

Light is an electromagnetic wave that is actually comprised of discrete packets of energy.

New York City in 1933 boasted many skyscrapers, but only one had an eighty-sixth floor. In our world the eighty-sixth floor of the Empire State Building is dedicated to the Observatory deck, but in the world of the pulps, this entire floor was rented to one man, who made it his residential home, complete with an extensive library and advanced chemical, medical, and electronic laboratories. This man, who excelled in all pursuits intellectual and physical, was frequently joined by his five close associates, each an expert in a different field of the practical and mechanical arts, such as chemistry, law, electronics, engineering, and archeology, on adventures that spanned the globe. The leader of this team, not content to rely solely on his amazing mental capabilities and his imposing physical prowess, would also employ a host of seemingly miraculous inventions and gadgets. Many of these exotic devices would not be realized in our world until years later, when nonfictional scientists and engineers had mastered the principles of quantum mechanics I've described, and managed to catch up to the achievements of one of pulp fiction's greatest heroes, Clark Savage, Jr. Though he had the equivalent of several Ph.D.s, owing to his M.D. from Johns Hopkins and several years studying brain surgery and neurology in Vienna, his friends and the public knew him as "Doc."

Doc Savage's adventures were described in the pulp magazine title that bore his name, and his first story, *The Man of Bronze*, was published in March 1933, written by Lester Dent. Before the year was out, *Doc Savage* would be one of the top-selling pulps on the newsstand. Dent would go on to write 160 more full-length Doc Savage novels over the next sixteen years, at a pace of nearly one a month.* Even at the pay rate of a penny a word, his writing income enabled Dent and his wife to live a life of personal adventure and travel that would inform his fictional tales. Doc Savage and his team would often travel the high seas in one of Doc's yachts or his personal submarine, battling modern-day pirates or exploring an island where dinosaurs still walked the Earth. Meanwhile, Dent and his wife lived for several years on a forty-foot schooner, traveling along the eastern seaboard, fishing and diving for buried treasure in the Caribbean by day and writing pulp adventures by night. Dent was a licensed pilot and radio operator, climbed mountains, prospected for gold in Death Valley, was a vast storehouse of obscure information, and was elected a member of the Explorers Club.

Dent's most famous literary creation would serve as the inspiration for Superman and Batman (Doc would retreat to an arctic sanctuary to develop new inventions that he called his Fortress of Solitude, and he carried many of his crime-fighting gadgets in a utility vest), James Bond and the Man from U.N.C.L.E. (Doc's tie and jacket buttons hid the chemical ingredients of thermite and his car could produce a smokescreen to blind pursuers), and Marvel Comics' Fantastic Four (the comic-book superhero foursome also lived in a skyscraper headquarters, and the friendly bickering between two of Doc's teammates presaged the relationship between the Thing and the Human Torch), and even *Star Trek*'s Mr. Spock (Doc could incapacitate foes by pinching certain nerves in their neck).

Doc's gadgets were similarly ahead of his time. In 1934 Doc employs a version of radar, long before its debut in World War II. (According to Dent, a reference to radar in a 1943 Savage novel was censored by the military immediately prior to publication, requiring him to scramble to come up with an alternative plot

---

* There were a few Doc Savage novels published in the 1970s and 1990s that were credited to Dent posthumously, but his main run on the pulp magazine ended in July 1949.

device).* Doc Savage employed shark repellant and colored dyes to mark a pilot's location when forced to eject over the ocean a good ten years before the navy would adopt these innovations. He invented a small tracking device that, when affixed to an automobile, would transmit a radio signal, enabling the car's position to be monitored from a remote location. And one of Doc's inventions—ultraviolet writing—employed in his first pulp adventure makes use of the same quantum mechanical principles that underlie the laser.

In 1933's *The Man of Bronze*, Doc and his team of adventurers search their quarters on the eighty-sixth floor for a message from Doc's recently deceased father. Knowing that his father would often leave him missives using a form of invisible writing, Doc brings out a small metal box that resembles a magic lantern. Showing the interior of the mechanism to Long Tom, the group's electrical expert, Doc tests his companion, asking him whether he recognizes the device. "Of course. [. . .] That is a lamp for making ultraviolet rays, or what is commonly called black light. The rays are invisible to the human eye, since . . . [their wavelengths] are shorter than ordinary light." Long Tom then points out that while we may not see in the ultraviolet, many common substances, such as quinine and Vaseline, fluoresce when so illuminated. When they shine this ultraviolet light on a window in Doc's office, sure enough, a message from his father is revealed in glowing blue letters, directing them to the hiding place where they would find important papers that would in turn send them on a perilous journey to the fictional Central American nation of Hidalgo. The mechanism by which Doc and his father, and in later pulp adventures Doc and his teammates, communicate through ultraviolet writing relies on the variation in transition rates for quantized levels.

We have seen that electrons bound in atoms are constrained to particular energy levels. A consequence of this discreteness is that the atoms can absorb or lose energy only when it enables transitions between these allowed energy states (we will neglect

---

* The government should not have bothered—in 1911 *Modern Electronics* published "Ralph 124C 41+," a science fiction story by Hugo Gernsback (who would go on to found *Amazing Stories*) that featured a fairly accurate description of radar, long before the term was coined.

transitions of protons or neutrons within the nucleus, as these energy scales are in the gamma-ray range, and we are interested now in transitions in the visible portion of the electromagnetic spectrum). Any energy interacting with the atom, in the form of a light photon or a collision with another electron or atom, will not induce an electronic transition if the change in energy does not correspond to the difference between two energy levels.

In our analogy of students in a classroom, where the rows of seats represent allowed energy levels, students may be promoted from their original seats at the front of the room to empty seats near the rear of the lecture hall. However, students are not allowed to stand between rows and may change their seats only if the energy they absorb takes them exactly from one row to another (and if the seat they are moving to is unoccupied). When an atom relaxes from some high-energy state back to the ground state, it similarly may do so only by emitting a photon whose energy is equal to the difference between the starting and final energy levels. That is, only electronic transitions that satisfy the principle of conservation of energy are allowed. This accounts for the discrete-line spectrum, with only a very select number of wavelengths observed (see Figure 13 in Chapter 5) when an atom is placed in a high-temperature environment. Different elements will have their allowed quantum levels at different energies, so that the spacing between levels, and hence the frequency of the light emitted when the electron moves between states, will differ.

Just because an electron can make a jump between two quantized energy levels does not determine how fast or slow such a transition may be. For a collection of atoms, the light will be brighter for those transitions for which the probability of a jump is higher. Some lines will be present, but very faint, as the probability of a transition occurring at any given moment might be very low. One of the great successes of the quantum theory is that it actually makes predictions of the transition rates, that is, the probability per second that an atom with an electron in an excited state would drop down to a lower energy state, emitting a photon in the process. Thus, the quantum theory correctly predicts not only what wavelengths will be observed for a given atom, but even how bright the lines will be.

What determines these transition rates is fairly complicated and depends on details of the wave functions for the initial and final states. The important point is that quantum mechanics is able to account for the following: (1) the fact that electrons in atoms may have only certain energies, (2) the fact that only certain transitions between allowed states are possible, and (3) the probability per second of a given transition occurring. That is, the theory can explain why only discrete lines rather than continuous spectra will be observed for the light emitted by an atom, as well as predicting the wavelengths of the line spectrum and the intensity of the lines, all in excellent agreement with experimental observations. We now know enough about how atoms interact with light to explain two of the most important inventions of the twentieth century: lasers and glow-in-the-dark action figures!*

Let's first consider glow-in-the-dark materials. Each atom in the solid has a highest occupied energy level (as in Figure 31), and when a trillion trillion of these atoms are collected, all of these "seats" broaden into an auditorium of quantum states, as illustrated in Figure 34. In Chapter 12 we saw that, thanks to the Pauli exclusion principle, each seat is actually a "love seat" in which two electrons can sit, if they have opposite spins (one with $+\hbar/2$ and the other with $-\hbar/2$). The trillion trillion "seats" in this "ground-state auditorium" can therefore accommodate two trillion trillion electrons.

If the atoms in the solid form bonds by keeping their electrons in "boxes," as in the case of the carbon-carbon bonds in diamond (Figure 32), then every love seat in the auditorium has two electrons, and the auditorium is completely filled (Figure 34a). The electron thus has to move to a higher energy (the next available empty quantum state) in order to find a vacant level. All of these higher energy states will also broaden into an "auditorium" of seats. Atoms that form solids similar to diamond can be considered to have an orchestra of seats, all of which are completely filled, and a higher-energy balcony with an equal number of seats, which are

---

* I'm sure that there was a point in time for nearly all readers when glow-in-the-dark materials seemed to be the greatest technological invention in the history of the universe.

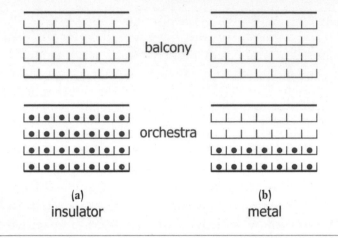

Figure 34: *Sketch of the band of quantum states from the highest energy occupied levels in a solid and the band formed from the next highest energy available quantum states. In an insulator (a) the lower band is analogous to a completely filled orchestra in an auditorium, where there is an energy gap separating the electrons in the lower band from the band of empty states (the balcony). The second figure (b) shows a situation where the lower orchestra is only half-filled and the electrons have ready access to empty seats—which describes a metal.*

all empty.* When a current flows in a solid in response to an applied voltage, the electrons gain kinetic energy, but this cannot happen if there are no unoccupied higher energy states accessible to the electrons. Consequently, only those electrons promoted to the balcony, by either heat or light, will be able to participate in an electrical current, moving along the newly available empty seats. Diamond is an electrical insulator because normally there are too few electrons in the balcony to provide an appreciable current.

In contrast, in metals the ground-state electrons are localized in "momentum space," and the orchestra that can seat *two* trillion trillion electrons is occupied by only *one* trillion trillion electrons. There are therefore many empty seats in the half-filled orchestra, as sketched in Figure 34b, and it is easy for the electrons to move from seat to seat when carrying an electrical current.

To construct a "glow-in-the-dark" nonmetallic solid, we need a filled orchestra, an empty balcony, and a "mezzanine" of seats,

---

* The lower energy seats that are filled for the individual atoms (Figure 31 c,d,e) also broaden into their own filled auditoriums that do not play a major role in the solid's properties.

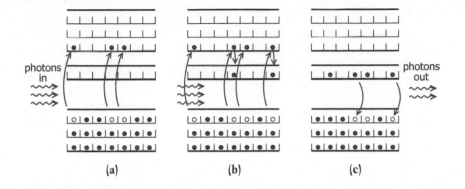

(a)   (b)   (c)

Figure 35: *Sketch of the band structure of a fluorescing solid, represented by a filled orchestra, an empty balcony at a high energy, and an unoccupied mezzanine level at a slightly lower energy than the bottom of the balcony. When the solid is illuminated with white light, electrons are easily promoted from the orchestra to the balcony, and photons are emitted when the electrons fall back into the lower level. Occasionally an electron will wind up in the mezzanine level, from which the transition rate to the orchestra is low. When the light exposure is stopped, these charges trapped in the mezzanine will eventually drop back into empty spots in the orchestra, emitting slightly lower energy photons in the process. In this way the material will give off light after being illuminated—that is, it will glow in the dark.*

also unoccupied, just below the balcony (sketched in Figure 35). Let's assume, for the sake of argument, that blue light is required to promote an electron from the orchestra to the balcony, but the mezzanine can be filled using lower-energy green light. The energy separation between the balcony and the mezzanine is in the infrared portion of the electromagnetic spectrum. These mezzanine seats may arise from a different element that is incorporated into the solid.

Now assume that the transition rate from the orchestra to the balcony is high. This means that it is easy to promote the electron up from the filled lower auditorium to the balcony, and once up in these states, the electron quickly falls back to the orchestra. The mezzanine is different—it has a very low transition probability, so that it is very hard to promote an electron from the orchestra into these levels. Once in the mezzanine, the electron has a very low probability of dropping back to the lowest energy state—it will thus sit in this state for a long time before dropping down.

Now, what will happen if we shine white light on this solid?

White light is comprised of all visible colors in equal intensities. Due to the discrete nature of the quantized energy levels, the atom will ignore all colors except for the blue and the green (let's not worry about the finite energy width of the orchestra and balcony for now). The blue light will be readily absorbed, as the transition rate for the orchestra to the balcony is high. Of course—easy come, easy go—and the electron in the balcony also has a high probability of dropping back down to the orchestra (in either its original seat or an empty seat created when another electron was promoted into the balcony), emitting a blue light photon as it does so (Figure 35a). For the most part this cycle continues—orchestra absorbs blue light, promoting electron to balcony; electron then releases another blue photon when falling back to the lower energy state. Occasionally, if we do this enough times, a seat in the mezzanine level becomes occupied, either by an electron being directly promoted from the orchestra to this level (just because the probability is low doesn't mean it won't happen if we try enough times) or possibly from the electron in the balcony dropping down into the lower-energy mezzanine instead of falling back to the orchestra (Figure 35b). We would not notice the infrared light emitted when the electron went from the balcony to the mezzanine unless we had specific detectors sensitive to this portion of the spectrum (alternatively, the electron can emit thermal energy as it moves from the balcony to the mezzanine). Once in the mezzanine, the electron will stay there until (1) an infrared photon excites it back into the balcony (not likely, as there is very little infrared light of the necessary energy in the white light source I am using); or (2) the electron drops back to an empty seat in the orchestra, emitting a green-light photon in the process (which can happen but has a low transition probability).

So, as we expose this solid to white light, blue light is absorbed and we get blue light back, but eventually the solid ends up with electrons sitting in the mezzanine, leaving unoccupied seats in the orchestra. Now the light is turned off. All the electrons that are still up in the balcony rapidly drop down into the empty orchestra seats, and then as time goes on, the electrons in the mezzanine seats also fall back to the orchestra (Figure 35c), emitting photons as they do, even if the solid is now in a completely darkened room, glowing in the dark! Eventually, as the number of electrons in the mezzanine decreases, the light emitted by the solid becomes dimmer and dimmer, until it is recharged with another prolonged exposure to white light. From such simple quantum mechanical phenomena are totally awesome toys made.

Doc Savage's invisible writing must employ an "ink" for which the separation between the orchestra and the balcony is in the far ultraviolet portion of the spectrum, while the spacing between the mezzanine level and the filled orchestra corresponds to blue light. Doc used the "black-light" lamp that emits ultraviolet light to promote electrons to the balcony, which then subsequently charge up the mezzanine. From the fact, as described in the pulp adventure, that the blue writing rapidly fades, we can assume that the electrons do not stay in the mezzanine level for more than a few seconds. The intensity of ultraviolet light in the Planck spectrum for sunlight is apparently too weak to charge up these states, which is why Doc needed to use the "black-light" lamp.

The energy separation between the balcony level and what we have termed mezzanine states, and how long electrons will remain in these states in the dark, depends on the particular elements that one introduces into the solid to produce these long-lived states. One does not need to use ultraviolet or visible light to promote electrons into these levels—any source of energy that can excite electrons from the orchestra to the balcony states can work.

Back in the 1950s, the hands of some alarm clocks were painted with radium, and the continuous emission of alpha particles would provide the energy necessary to keep the balcony in the phosphor material occupied, thereby enabling the hands to glow in the dark. When the radium emits an alpha particle, the nucleus converts into radon, which is also radioactive. Eventually the materials for glow-in-the-dark alarm clocks were replaced with less toxic substances. Nevertheless, radioactive materials, and their ability to emit sources of energy at a uniform rate, are hard to give up. Smoke alarms use a radioactive isotope to create a beam of particles, and an alarm is triggered when this beam is obscured from its detector by smoke or haze. Certain wristwatches with glow-in-the-dark faces have replaced radium as the radioactive element that excites the phosphorescent material with high-energy electrons from the decay of tritium as the source of external energy. Most diners are likely relieved that Fiestaware dishes no longer employ uranium oxide in their bright orange-red glaze, as they did back in the 1930s. The shine on modern Fiestaware dinner plates may be not quite as bright, but it is much safer.

CHAPTER FIFTEEN

Death Rays and DVDs

The popularity of the Buck Rogers newspaper strip led to a similarly successful radio serial program, and in 1934 a competing strip featuring the adventures of Flash Gordon was introduced. By the mid-1930s the demand for Buck Rogers– and Flash Gordon–inspired toy ray guns was so high that the Daisy Manufacturing Company, which had the license to create stamped-metal versions of Buck's XZ-31 Rocket Pistol, ran out of both steel and cardboard boxes. Given the association of ray guns with the future conquest of space, perhaps it is not surprising that in 1960, when the development of the laser was announced, the first thing the public wanted to know was whether science had at last delivered the long-anticipated "death ray."

A patent for a laser, capable of projecting a high-intensity beam of visible light, designed by Charles H. Townes and Arthur L. Schawlow at Bell Labs, was filed in 1958, and in 1960 Theodore H. Maiman at Hughes Research Laboratory in California successfully constructed the first working device. At his press conference in 1960, Maiman was peppered with journalists' questions about whether he had in fact invented a death ray. When speaking to the public, scientists from Bell Labs were instructed by management to deflect any questions concerning using the laser as a lethal weapon and took pains to avoid saying anything that might be misconstrued or misquoted. Yeah, good luck with that. In 1961, the report in the *Detroit News* of a lecture by a Bell Labs scientist involved in their laser program prominently featured "Death Ray" as the invention's first potential application. Four years after Maiman's announcement, in 1964's MGM film *Goldfinger* James Bond is threatened with a slow, painful death while strapped to a table. The circular buzz saw of the 1959 novel was replaced in the movie with a high-power industrial laser, its beam slowly moving along the length of the table on a trajectory intended to bisect Agent 007.

The physics of the laser is essentially that of a glow-in-the-dark solid. Depending on their chemical composition and material properties, lasers can emit not just green light, but red, green, blue, ultraviolet, or infrared photons. The two big differences between lasers and glow-in-the-dark solids is that in lasers, the mezzanine levels are nearly completely occupied with electrons, and, more important, when the electrons in the mezzanine level drop down to the ground state, they all do so at the same time.

How can one ensure that all the electrons residing in the laser levels will choose to drop down to the ground state, emitting photons, simultaneously? Consider the auditorium analogy for a solid, shown in Figure 35.* I use essentially the same argument as for the glow-in-the-dark situation from the last chapter. Electrons from the filled orchestra level are promoted up to the balcony by, for example, the absorption of light, or an electrical current. The electrons excited up into the balcony leave behind empty seats in the orchestra. The transition rate is high for electrons to go from the orchestra to the balcony, and it is similarly easy for these electrons to drop back down into the orchestra, emitting light as they do so. Occasionally, an electron will not fall from the balcony to the orchestra, but into a mezzanine seat instead. The transition rate into or out of these mezzanine levels is very low, so once the electron is in one of these quantum states it will stay there for quite some time. If electrons can be excited up to the balcony, and from there to the mezzanine, faster than they spontaneously drop

---

* Note that a laser does not need to be a solid—the helium-neon laser pointers used by public speakers and lecturers employ a mixture of two inert gases to generate laser light. For simplicity, we'll stick with solid-state lasers, but our arguments hold just as well for gas lasers.

down from the mezzanine level back to the orchestra, then we can obtain a situation where we have nearly as many electrons in the mezzanine level as in the orchestra.

We are now ready for some laser action, as shown in Figure 36. There are two ways that an electron in the mezzanine band can return to an empty seat in the orchestra—it can fall or it can be pushed. The transition rate for an electron to spontaneously fall from the mezzanine to the orchestra can be, for some materials, up to a hundred million times slower than for the electron to move from the balcony to the orchestra. This was why we needed to go through the balcony levels in order to populate this intermediate energy band. What could push an electron down to the orchestra? Light.

During the transition from the mezzanine to the orchestra, the electron's wave function can be expressed as the overlap of the orchestra and mezzanine quantum states. During this process the electron's average location may be considered to oscillate between its value for each state. An oscillating electric charge emits electromagnetic waves at the frequency of vibration. A formal quantum mechanical analysis of this process finds that the energy emitted is in a discrete packet of energy (that is, a photon) whose energy corresponds to the energy difference between the mezzanine and orchestra levels.*

Once a photon is emitted, this quantum of the electromagnetic wave can induce oscillations in another electron up in the mezzanine level, making it easier for this second electron to jump down into the orchestra, emitting its own light quantum in the process. This second photon can stimulate another electron to make the transition, generating yet another photon with an energy given by the separation of the mezzanine and orchestra bands. In this way a cascade of falling electrons, each induced (pushed) by the oscillating electric field of a light quantum, results. One photon in therefore leads to potentially trillions of photons out, all with ex-

---

* Experts will note that the above argument applies to electric dipole transitions, but not to those involving magnetic dipole or electric quadrupole transitions. These are typically ten thousand to a million times less likely than electric dipole transitions. For the nonexperts—nothing to see here; move along!

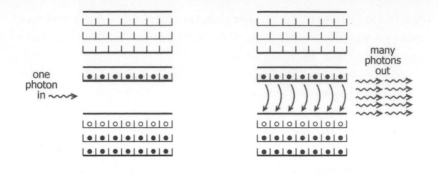

Figure 36: *The auditorium model from the last chapter, only now the occupation of the mezzanine level is quite high. A single photon can stimulate an electron in the mezzanine to drop down to an empty seat in the orchestra, emitting a photon in the process. This photon can in turn induce another electron to make this transition, with the net effect that a very large number of electrons may be stimulated into dropping down to the lower energy band, all emitting identical energy photons. This procedure is the basic physical mechanism underlying the laser.*

actly the same energy, emitted all at the same time. As the photons are fast, as in speed-of-light fast, there is no noticeable time lag between the first electron falling from the mezzanine and the trillions of electrons stimulated by other photons. The device produces *l*ight *a*mplification by *s*timulated *e*mission of *r*adiation and is called a "laser" for short.

Of course, if I want this stimulated emission of light to occur more than once, I have to continue to excite electrons up to the balcony level, so that I can maintain the population inversion of electrons in the mezzanine. Thus, it will take a great deal of energy to run the laser. The more photons that I want to be emitted per second, the more energy I have to expend maintaining the occupancy of the mezzanine level. A laser pointer used in a lecture presentation is relatively low intensity and can be run from two AA or AAA batteries, while the high-power versions used in industrial-laser cutting procedures require a thousand Watts of power, enough energy to run a standard household.

Lasers make use of the fact that the emitted light is coherent (that is, all the light waves are in phase with one another, as in the constructive interference example from Chapter 2, Figure 4). The material that is being stimulated to emit photons is typically housed in a long cylinder, both ends of which are mirrored, with one end having a small hole for light to escape. Consequently all the walls of the auditorium reflect photons, and only those light quanta moving in exactly the right direction toward the single exit will depart the hall.* Those photons that do not leave the chamber will bounce back and forth, inducing more transitions from the mezzanine to the lower level. The laser light thus forms a tightly focused beam, and as the photons are in phase, they will exhibit minimal spreading upon leaving the laser cavity. Laser light is therefore invisible unless you look directly at the aperture of the laser cylinder, unlike incandescent lightbulbs, from which the illumination spreads out uniformly in all directions. We can see light from an incandescent bulb regardless of where we are looking, but in a sense these photons' energies are wasted, as light is hitting objects I don't care about seeing. The laser beam can be seen only if it reflects off a surface. If there is no dust or particulates in the air to scatter the laser beam, the only way to see it is when it gets to where it is going. A tight, narrow laser beam, sent out from a lab on Earth, was measured to have broadened out to a width of only about two miles after traveling 240,000 miles to the moon.

Thanks to the quantization of energy levels, when the electrons drop from the mezzanine to the lower-energy orchestra in response to the photon stimulation, they will all emit light of exactly the same energy. The light from a laser will thus be of a single frequency, that is, one color, with remarkably small variations. An efficient mechanism to generate red laser light is to use a mixture of two gases, helium and neon. Both of these elements have completely filled outer quantum levels (as shown in Figure 31b) and are thus chemically inert—they do not lower their energy by forming any type of chemical bond. When an electron beam is passed through this gas mixture, the kinetic energy of the electron current can be transferred when it collides with a helium atom. An electron in the helium atom is excited from the ground state to an "excited state"—which we have been terming the balcony level. The spacing of their quantum levels is such that when the helium atom with its electron in the higher-energy state collides with a

---

* The operation of a real laser cavity is a bit more complicated than this.

neon atom, it promotes an electron into a very long-lived excited state in the neon atom that acts as the mezzanine level. When light of the necessary frequency stimulates the neon atoms, they drop back to their ground state, emitting red photons.

By using electrically charged (that is, ionized) argon gas instead of a helium-neon mixture, green light can be produced. Using semiconducting diodes (much more on this in the next chapter), one can dispense with the gases and construct a completely solid-state laser, capable of producing red, green, or even blue light. Red light has a lower energy, of 1.9 electron Volts, and longer wavelength (about 650 nanometers) compared to blue light's photon energy of 2.6 electron Volts and a wavelength of 475 nanometers. The difference in wavelength may not seem like much, but it makes a big difference in your DVD player.

Anyone who has closely examined an old-style newspaper photograph, composed of a series of black and white dots, understands that the information contained in an image may be relayed via a series of pixels. Digital versatile discs (DVDs) and compact discs (CDs) encode images and sound or just sound, respectively, through a set of instructions for either a video display or audio system. Pixels are binary, in that they have just two states: on or off, bright or dark. All digital data representation basically involves strings of "ons" and "offs," often referred to as "ones" or "zeros."

The development of inexpensive, compact solid-state lasers enables one to "read" the storage of these ones and zeros on a disc. A laser is bounced off the shiny side of the disc, and the reflected light is detected by an optical sensor. If the surface of the disc is smooth, then the laser light, which travels in a straight line, will be reflected directly onto the optical detector, and that location on the disc will be recorded as being a bright spot. If the laser light falls on a region of the disc that is distorted (for example, at the edge of a little pit gouged into the disc or a bump protruding from the surface), then the light will scatter in some random direction and not be reflected onto the optical detector. The detector will thus indicate a dark spot at this location of the disc. Calling the bright spot a "zero" and the dark spot a "one," we can store and transmit digital information.

Moving the laser along the disc, one can record the sequence of smooth and rough regions and translate that into ones and zeros, which in turn can be decoded to make beautiful music. Actually, it's easier to keep the laser fixed and move the disc underneath it (rotating the disc at high speed—typically at several hundred revolutions per minute) as the laser spot is moved from the center of the disc to its outer edge. The higher the density of ones and zeros (that is, the more bits of information in a given length), the higher the resolution of the video or audio signal. Here is where innovations in laser technology, thanks to quantum mechanics, have had a real impact on consumer entertainment technology.

If you wish to paint a two-inch-high statuette of an Orc (to take a random geeky example), you do not use the same large brush you would use for painting your house (assuming you are interested in doing more than just glopping a single color of paint on the figure). In order to apply different colors over the small details on the tiny character, you will need a very fine brush that would make house painting tedious but is well suited for the detailed work on the statuette. When light is used as a probe, the wavelength plays the same role as the fineness of the brush's bristles. One cannot use a wave to detect features smaller than the spacing between the peaks or troughs of the wave.

This is why optical microscopes, using visible light whose wavelengths are on the order of several hundred nanometers, are not able to let us see viruses or other nanometer-scale objects, regardless of the focusing. To "see" such small-scale structures, either you need light with a wavelength on the order of nanometers or smaller (such as high-energy X-rays, which lead to the necessity to develop X-ray lenses and focusing procedures) or you can employ electrons. The de Broglie wavelength of electrons can be adjusted by varying the momentum, which is easy to control by changing the magnitude of the accelerating voltage acting on the electron beam, and a series of charged plates can focus the electron beam. Detection of the current either reflected from a surface or transmitted through a thin sample can thereby provide "images" with atomic-scale resolution, and these electron microscopes are another example of quantum mechanics in action.

In the early days of compact disc storage media, only infrared solid-state diode lasers were available. The wavelength of infrared light is fairly long, so the density of bits (bits per area) was low. As the size of the disc was fixed, this meant that the spacing between pits on the disc had to be relatively large, and a typical disc could hold roughly 600–800 million bits. With fewer ones and zeros available, these pioneering compact discs could store enough information for music but not enough for high-quality video.* With the fabrication of visible-light red solid-state lasers, the wavelength of the light decreased and the number of bits that could be squeezed on a disc similarly increased, up to approximately five billion bits. These digital discs were highly versatile (hence the name DVD), as they could encode both images and music. With the recent innovation of relatively inexpensive blue-light solid-state lasers, the density of bits can be increased even further. Now the same movie can be stored using a much greater number of pixels per inch, and these high-definition Blu-ray DVD players (where "Blu" stands for blue) can bring theater-quality video to the home.

How do the pits get on the DVD disc in the first place? With another laser. Readers of *Dr. Solar—Man of the Atom* # 16 in 1966 were treated to a feature page after the regular story, divulging "Secrets of Atom Valley." One such page discussed the "Birth of the Death Ray," which in comics at the time consisted of a laser mounted on a rifle. The concentrated beam of photons emanating from a laser can indeed do great damage, depending on the surface it illuminates. The light carries energy, and when the material absorbs this light, it must have a way to dissipate the excess energy per atom provided by the laser. "Temperature" is a bookkeeping device used in physics to keep track of the average energy per atom in a system. If the material cannot reemit the energy absorbed as light, then it must do so as atomic vibrations. That is, the material will heat up due to the application of the laser light, and if the power density (that is, the number of absorbed photons per area per second) is large enough, the material can be heated by the laser faster than the excess heat can be transferred to the rest of the solid. In that case the atoms may be shaken so violently that they break

---

* Old-timers may recall an early ancestor of the DVD—laser discs, which were twelve inches in diameter in order to have sufficient room for the relatively low-density bits encoding a video image that could be read with a laser.

the bonds holding them in the material, and either melt or vaporize. The power of early lasers in the 1960s was characterized by the number of Gillette razor blades they could melt through. Laser ablation, where a laser beam evaporates a material, creating a vapor of a substance that is ordinarily a solid, is used in research laboratories to synthesize novel semiconducting materials, when the resulting vapor condenses onto a substrate or reacts with another chemical.

When writing information on DVDs and CDs, say, in the CD/DVD burner in some home computers, the laser need not vaporize the disc. Rather, either it induces a chemical change in a dye that coats the disc, darkening it so that it is no longer reflective, or it can melt the material under the laser spot. When rapidly cooled, instead of being a smooth, uniform surface, the newly melted region will be rough and will ably serve as a "pit" that will scatter a second laser beam in the CD or DVD player. For commercially

Figure 37: An "educational" page from Dr. Solar—Man of the Atom *# 16 in 1966, showing how lasers can be used for good or evil.*

manufactured CDs and DVDs, a laser is used to cut a master disc, which is then used to stamp out multiple copies that contain the encoded information.

The trouble with using lasers as "death rays" is that it is difficult to achieve the necessary power density needed to wreak any significant mayhem. To locally melt a small region on a DVD disc, one must supply a significant amount of energy in a short amount of time—faster than the energy can be transferred to the rest of the material. The issue is thus the rate at which the energy can be delivered, which in physics is termed the "power." One could construct a laser capable of melting large holes in the steel plating of tanks, but the laser would be as large as a desktop—not counting the required power supply.

The last panel of the Dr. Solar informational page in Figure 37 alludes to the laser's potential for healing, as well as harm. This was anticipated in comic books as well. In "A Matter of Light and Death" in 1979's Action # 491, Superman removes the thick cataracts that have blinded a companion by using his focused heat vision. First Superman takes two lumps of coal and squeezes them until they form large, perfect diamonds. This is harder than you think—as Superman muses while compressing the coal, "Transforming carbon from its crude coal form to its purest state is no easy trick . . . even for me! After all, it takes Mother Nature millions of years and just as many tons of underground pressure to produce even one raw diamond . . . let alone two!" As coal is fossilized peat moss, what happens to the impurities in the lumps Superman squeezes, such as sulfur, nitrogen, and other chemicals, which are present in coal but not in an optically pure diamond, is not revealed. Holding these two large diamonds in front of his friend's eyes, he then uses his heat vision. As he performs the operation, the Man of Tomorrow thinks to himself, "These diamonds are filtering and concentrating my beams of heat vision into two super-laser beams—enabling me to do what man-made lasers couldn't—burn away those cataracts and restore his eyesight." (Good thing for readers that superheroes always narrate their actions in their heads!) Eight years later, Dr. Stephen Trokel would patent and perform the first non-superhero-enabled laser eye surgery, using an excimer laser that emits ultraviolet light (as opposed to Superman's heat vision, which presumably consists of infrared light) and was previously used to pattern semiconductor surfaces. While not a common method for treating cataracts, laser surgery for re-forming the cornea to correct myopia and other refractive vision processes is now quite common.

Did Doc Savage understand all of this when he communicated via invisible writing that could be read only under ultraviolet illumination, employing the same physics as glow-in-the-dark solids? Perhaps he didn't know all the details of how high-resolution DVD players work, but we need not wonder whether Doc was familiar with basic quantum mechanics. In 1936's Doc Savage adventure *The South Pole Terror,* Doc and his band of adventurers foil the elaborate scheme of a group of thieves and murderers who attempt to mine platinum from an Antarctic valley. The crooks are able to melt vast quantities of ice, and also kill interfering witnesses, using a strange heat ray, whose operation mystifies all but Doc. As he explains at the tale's conclusion: "It has long been known that the atmosphere layer around the earth stops a great many rays from the sun. Some of these rays are harmless, and others are believed capable of producing death or serious injury to the human body. [. . .] The particles of air, for instance, are made up, according to the Schroedinger theory, of atoms which in turn consist of pulsating spheres of electricity."

Doc had correctly surmised that his opponents had "an apparatus for changing the characteristics of a limited section of atmosphere above the earth to permit the entrance, through this atmospheric blanket, of the cosmic rays." Nine years before the Manhattan Project, Doc Savage was citing Schrödinger and fighting fiends who possessed a device that could open, at will, a hole in the ozone layer above Antarctica, demonstrating his mastery over both quantum physics and evildoers.

CHAPTER SIXTEEN

The One-Way Door

Matter is comprised of discrete particles that exhibit a wavelike nature.

Science fiction pulps and comic books from fifty years ago told of how, by the year 2000, robots would break free of their shackles of servitude and rebel against their human overlords. In order to be capable of such insubordination, these automatons must have electronic brains capable of independent thought and initiative. They must therefore possess very sophisticated computers that are able to go far beyond the mathematical calculations of the "difference engines" of the 1950s. Such powerful computers are closer to reality than the writers back then may have thought, thanks to scientists at Bell Labs who, in 1947, making use of the advances in our understanding of the solid state afforded by quantum mechanics, developed a novel device that would dramatically shrink the size and simultaneously expand the computing power of electronic brains—the transistor.

In the 1957 issue of DC comics *Showcase* # 7, the Challengers of the Unknown crossed swords with a sophisticated computer atop a giant robot body. The Challengers are four adventurers—a test pilot, a champion wrestler and explorer, a professor and deep-sea-diving expert, and a mountain climber and circus daredevil—who are the sole passengers on a doomed cross-country flight. Crashing in a freak storm, the plane is completely destroyed, yet the four passengers walk away from the carnage unharmed. Realizing that they are "living on borrowed time," they devote their lives to adventure, repeatedly throwing themselves in harm's way as they

Figure 38: *Panels from* Showcase # 7, *where the Challengers of the Unknown discover that "ULTIVAC Is Loose!" Hesse, a German scientist captured by the Allies, gives physics lessons to his cellmate, bank robber Floyd Barker. Upon their release, the pair design and construct a "new type of calculating machine"—* ULTIVAC!

thwart alien invaders, mad scientists, and undersea monsters. Frankly, my response to the premise of this team's origin would be quite different—if I were to survive a horrific car crash, for example, I doubt I would then jump out of airplanes without a parachute or juggle nitroglycerin reasoning that as I could have died in the traffic incident, I am now able to take additional insane risks. But luckily for comic book readers, the Challengers' attitude toward danger differed from my own, as not only were their tales exciting in their own right, but the foursome served as one of the models for Marvel Comics' 1961 superteam, the Fantastic Four.

The Challengers' challenge in *Showcase # 7* is ULTIVAC, a fifty-foot-tall robot capable of independent thought. ULTIVAC is constructed by Felix Hesse, a German scientist who was captured by the Allies at the end of World War II and sent to prison as a war criminal. Hesse is assisted by Floyd Barker, a bank robber he meets in prison. They pass the time with physics lessons—the scientist teaching Barker ("All it takes is some study!" says the bank robber). Not too long after being released, they design and construct a giant calculating machine, as shown in Figure 38. The scientist remarks that ULTIVAC must be enormous "to do all the things we want it to do! This is going to be the greatest calculator of all time!" Apparently, as later revealed in the story, their get-rich scheme involves exhibiting their creation in Yankee Stadium, charging admission to see "two tons of steel . . . that thinks and talks like a man!" ULTIVAC rebels against this public display and flees, joined by Dr. June Robbins, a scientist who convinces ULTIVAC that humans and computers can be friends. Addressing an assembly of politicians, scientists, and national leaders, ULTIVAC promises, "I am willing to apply my power to the cause of helping mankind—if humanity meets me halfway!" However, rather than hand over to the government what he imagines to be a source of great wealth, the German scientist who built ULTIVAC damages him mortally. Emergency repairs are effected, and in the final panel we see that ULTIVAC is now a stationary calculating machine. As shown in

Figure 39: *Final panel from* Showcase # 7, *where Dr. June Robbins describes to the Challengers of the Unknown the final disposition of ULTIVAC—while "just a stationary calculating machine," it is "still contributing much to man's knowledge."*

Figure 39, Dr. Robbins has the last word, telling the Challengers, "The spark that made ULTIVAC think like a man is gone! But as a pure machine, he is still contributing much to man's knowledge!"

By 1957, computers had indeed begun contributing much to humanity's knowledge, helping us with complex tasks. In 1946, scientists at the University of Pennsylvania constructed the first electronic computer, called ENIAC, for Electronic Numerical Integrator and Computer. It was more than eighty feet long and weighed nearly fifty-four thousand pounds. As the semiconductor industry did not yet exist, ENIAC employed vacuum tubes—nearly 17,500 of them—and more than seven thousand crystal diodes. It was owned by the U.S. military, and its first calculations were for the hydrogen bomb project. In 1951, the same scientists who built ENIAC, now working for Remington Rand (which would become Sperry Rand), constructed UNIVAC, a UNIVersal Automatic Computer that consisted of more than five thousand vacuum tubes and was capable of performing nearly two thousand calculations per second. This computer sold for more than $125,000 in the early 1950s to the military or large corporations and was used by CBS-TV to predict Dwight D. Eisenhower's victory in his run for the presidency in 1952. UNIVAC was unable to walk or fight jet planes, but it was about the size of ULTIVAC's electronic brain, as shown in Figure 39. Absent the semiconductor revolution, increasing the computing power of such devices entailed using more and more vacuum tubes and complex wiring, and as mentioned in the introduction, only a few large companies or the government would have the resources to purchase such machines.

The groundwork for the dramatic change that would reverse this trend, leading to smaller, yet more powerful computers, began in 1939, when a Bell Labs scientist, Russell Ohl, invented the semiconductor diode. We now know enough quantum mechanics to understand how this device and its big brother—the semiconductor transistor—work, and why many believe them to be the most important inventions of the twentieth century.

The first thing we need to address is the definition of a "semiconductor." We discussed two types of materials in Section 4—metals and insulators. Metals satisfy the Pauli exclusion principle by allowing each atom's "valence" electrons (those last few electrons not paired up in lower energy levels) to occupy distinct momentum states. The uncertainty in their momentum is small, and the corresponding uncertainty in their position is large—as these electrons can wander over the entire solid. At low temperatures there are many electrons available to carry an electrical current. Insulators satisfy the requirements of the Pauli principle by spatially restricting each atom's valence electrons, keeping them localized in bonds between the atoms, like the carbon-carbon bonds in diamond, sketched in Figure 32 in Chapter 12. At high temperatures, some of these electrons can be thermally excited to higher energy states (that is, from the orchestra to the balcony), where they can conduct electricity, but at low temperatures all the electrons stay locked within each atomic bond and the material is electrically insulating.

But what is a "low" temperature? Low compared to what? A convenient and natural temperature scale to compare "low" and "high" to would be room temperature. In this case, there is a third class of materials that are much better conductors of electricity at room temperature than insulators such as glass or wood, but much poorer conductors than metals such as silver or copper. These partway-conducting solids are termed "semiconductors."

Recall from our discussion in the previous chapter that a laser is a material with an orchestra of seats, all filled with electrons, separated from a balcony where all the seats are empty. Let us ignore for the time being the "mezzanine" we posited residing between the filled orchestra and the empty balcony (we'll get back to those states soon). In an insulator the energy separation between the orchestra and the balcony is typically five to ten electron Volts, well into the ultraviolet portion of the electromagnetic spectrum. Consequently, only light of this energy could promote an electron up into the balcony (like Doc Savage's "invisible writing" from Chapter 14). The intensity of this light is normally low, and at room temperature there isn't enough thermal energy from the atoms to promote a significant number of electrons to the empty balcony. Consequently, if a voltage is applied to an insulator, there is a negligible current at room temperature. In a semiconductor, the energy gap between the orchestra and the balcony is much smaller, usually one to three electron Volts. Visible light photons, of energy between 1.9 and 3.0 electron Volts, have sufficient energy to excite electrons up to the conducting balcony. Similarly, at room temperature, the thermal energy of the atoms is large enough to excite some electrons into the upper band. Of course, the larger the energy separation between the filled states and the empty band, the fewer electrons will be thermally excited up to the conducting balcony at room temperature.

Semiconductors make convenient light detectors, as the separation between the bands of filled and empty states corresponds to energies in the visible portion of the spectrum. A particular material will have an energy gap of, let's say, one electron Volt (which is in the infrared portion of the spectrum that our eyes cannot detect). Normally, in the dark some electrons will be thermally promoted to the empty conduction band, leaving behind empty seats in the orchestra. These missing seats are also able to conduct electricity, as when an electron moves from a filled seat to occupy the empty one, the unoccupied state migrates to where the electron had been, as sketched in Figure 40. These missing electrons, or "holes," in a filled band of seats act as "positive electrons" and are a unique aspect of the quantum mechanical nature of electrical conduction in solids. This process occurs in insulators as well, only then there are so few empty spots in the lower-energy orchestra, and so few electrons in the balcony, that the effect can be ignored. The electrons up in the balcony in the semiconductor will fall back into the empty seats in the orchestra, but then other electrons will also be thermally promoted up to the empty conducting band. So at any given moment there are a number of electrons and holes in this semiconductor that can carry current. The current will be very small compared to what an equivalent metal wire could accommodate, and a circuit with the semiconductor will look like it has an open switch in the dark. When I now shine light of energy one electron Volt or higher on this semiconductor, depending on the intensity of the light, I can excite many, many more electrons into the empty band, and leave many, many more holes in the filled band. The ability of the material to conduct electricity thereby increases dramatically. In the circuit it will look as if a switch has been closed, and the electronic device can now perform its intended operation.

And that's how quantum mechanics makes television remote

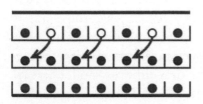

Figure 40: Sketch of nearly filled lower energy and nearly empty higher energy bands in a semiconductor. There will be some electrons promoted up to the "balcony" that can carry current (as they have easy access to higher energy quantum states, so they are able to gain kinetic energy and carry an electrical current). At the same time the vacant seats in the orchestra are also able to act as positive charge carriers, as other electrons slide over to fill the vacancy.

controls possible!* The remote control sends a beam of infrared light (invisible to our eyes) to your set. If you point the front edge of the device away from the set, the signal does not reach the photodetector and the setting remains unchanged (with certain models one is able to bounce the infrared beam off a wall and still have a sufficient intensity of photons reach the set to be detected). Once the light beam reaches the semiconductor and is absorbed, the conductance of the material increases and the circuit is closed. The infrared beam sent when you press a button on the remote control encodes information through a prearranged series of pulses (not unlike Morse code), and thus, different instructions can be transmitted to the set.

This is the same physics by which your smoke detector works. Some models use a beam of infrared light directed toward a photodetector. When the particulates in the smoke scatter the beam away from the detector, the circuit is broken and a secondary circuit sends current to the loud, high-pitched alarm. Other models employ a small amount of the radioactive isotope americium, which emits alpha particles when it decays. These alphas electrically charge the air in the immediate vicinity of the source, and the electrical conductivity of the charged air molecules is measured. Smoke particles trap these charges, and again, once the primary

---

* To the men of America—you're welcome!

circuit is broken, a secondary circuit sets off the alarm. From automatic doors that open when you approach, to street lights that turn on when darkness falls, we do not notice how often we employ semiconductors' ability to change their electrical properties dramatically when illuminated by light.

These photodetectors played a key role in a broadcast of *The Shadow* radio show back in 1938. The Shadow, who in reality is Lamont Cranston, wealthy man-about-town whose true identity is known only to his constant aide and companion, Margo Lane, has learned while in the Orient various mental powers that enable him to cloud men's minds. In *Death Stalks the Shadow*, a crooked lawyer, Peter Murdoch, sets a death trap for the Shadow using solid-state light sensors. When Lamont and Margo are out at a nightclub, they note a gimmicked door that opens whenever a waiter approaches. Lamont explains to his companion that the door is controlled by a photoelectric ray emitted by and detected by chromium fixtures on either side of the door, so that whenever the beam is broken, the door is opened. Lamont muses that such innovations pose a risk for him, as "the Shadow can hide himself from the human eye, Margo, but he has a physical being, and the photoelectric beam could detect his presence."

This is just the plan of Peter Murdoch, who hires an electrician to wire a sealed room with a steel door that will slam shut when a similar invisible beam ("You can't see it. The beam is infrared," explains the electrician) is broken when the Shadow enters the room. The death-room trap set, the electrician is murdered so that he cannot reveal Murdoch's plans. The Shadow does indeed enter the room, the steel door slams tight and is electrified, and poison gas is pumped into the room. Through this all the Shadow chuckles his low, sinister laugh. For he knows not only what evil lurks in the hearts of men, but also that in 1930s radio serials, even master criminals with law degrees are not very smart. To taunt his adversary, Murdoch has left the body of the electrician in the room with the Shadow. Removing a pair of pliers from the dead worker's overalls, our hero proceeds to disable the electricity in the room. The door no longer a threat, the Shadow escapes, captures Murdoch and his gang, and hands them off to Commissioner Weston and a promised cell on death row (the weed of crime bears bitter fruit, after all). Even infrared photodetectors are no match for . . . the Shadow!

But if this were the only advantage of semiconductors, the world we live in would not look that dramatically different from that of the 1930s. The real power of semiconductors is realized when different chemical impurities are added to the material, a process that goes by the technical term "doping." Consider Figure 41, featuring filled states, and the empty band of states at higher energy, likened to the filled orchestra and empty balcony in a concert hall. When discussing the physics of lasers, we introduced a "mezzanine" level, at a slightly lower energy than the balcony, which resulted from the addition of another chemical (typically phosphorus) to the material. In semiconductors there are two kinds of "mezzanines" that can be incorporated, depending on the specific chemical atoms added—those that are very close to the empty balcony and those that are just above the filled orchestra. If I manage my chemistry correctly, I can ensure that the benches right below the empty balcony have an electron in their normal configuration (Figure 41a). Then, even at room temperature, since there is only a very small gap in energy between the occupied bench and the empty balcony, nearly all the electrons will hop up to the balcony, and the holes they leave behind will be not in the orchestra, but in the seats in the upper mezzanine (Figure 41b).

Similarly, with a careful reading of the periodic table of the elements, a narrow band of seats (a "lounge," let's call it) can be placed just above the filled orchestra (Figure 41c). These lounge seats normally would be empty of electrons, depending on the chemistry of the added atom and the surrounding semiconductor material. An electron can then jump up from the filled orchestra, leaving a hole in the lower band without having to promote an electron up in the balcony (Figure 41d). The seats in the lower-energy lounge band, as well as the higher-energy mezzanine, are far enough apart from each other that it is hard for an electron or hole to move from seat to seat in these states. The mezzanine and lounge states are ineffective at carrying electrical current, but they can dramatically change the resistance of the surrounding semiconductor by easily adding either electrons to the balcony or holes in the orchestra. The first situation, with the mezzanine adding electrons to the

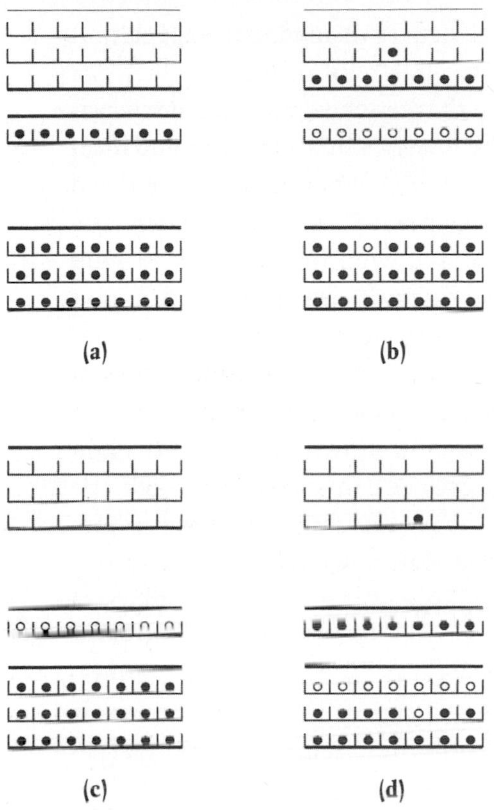

Figure 41: *Sketch of a semiconductor where impurity atoms are added, resulting in a mezzanine level beneath the balcony, which at low temperatures is normally filled with electrons (a) that are easily promoted at room temperature into the previously empty balcony (b). Alternatively, different chemicals can produce states right above the filled orchestra (c) that at low temperatures are normally empty of electrons. At room temperature electrons can be easily promoted from the orchestra to these lower "lounge" seats, leaving empty seats (holes) in the orchestra that are able to carry electrical current (d).*

balcony, is called an n-type semiconductor, since I have the net effect of adding mobile negatively charged electrons, while the second situation, with a low-energy lounge accepting electrons from the orchestra, leaving behind holes in the lower band, is termed a p-type semiconductor, as the current-carrying holes added are positively charged. As the atoms added to the material were previously electrically neutral, promoting an electron to the balcony from the mezzanine leaves behind a positively charged seat in these upper states, and accepting an electron into the lounge, leaving a mobile positively charged hole in the orchestra, makes the lounge seat negatively charged.

If we added either n-type impurities or p-type impurities to a semiconductor, then the number of electrons or holes would increase, with the effect that the semiconductor would be a better conductor of electricity. Of course, if all we wanted was a better conductor of electricity, then we could have used a metal. No, the real value of doping comes when we take two semiconductors, one that has only n-type impurities so that it has a lot of mobile electrons in the balcony and holes stuck on the benches, and another semiconductor with mobile holes in the nearly filled orchestra and electrons sitting on the benches near the filled band, and bring them together. If these two pieces were each a mile long, then we would expect that very far from the interface each material would look like a normal n-type or p-type semiconductor. But the junction between the two would be a different matter.

The n-type material has electrons in the balcony but no holes in the orchestra, while the p-type semiconductor has mobile holes in the orchestra but none in the balcony. As shown in Figure 42, when they are brought together, the electrons can move over from the n-type side to the p-type (and the holes can do the reverse), where they combine, disappearing from the material. That is, the electrons in the balcony can drop down into an empty seat in the orchestra (remember that the Pauli exclusion principle tells us that no two electrons can be in the same quantum state, so the electron can drop down in energy only if there is an empty space available to it), and it will be as if both an electron and a hole were removed from the material. But the positive charges in the seats in the mezzanine in the n-type solid and the negative charges in the seats in the lounge in the p-type material do not go away. As more and more mobile electrons fall into the mobile holes, more positive charges in the mezzanine in the n-type material and negative charges in the lounge in the p-type material accumulate, neither of which can move from one side to the other.

The net effect is to build up an electric field, from the positive charges in the n-type side to the negatively charged seats in the p-type material. Eventually this electric field is large enough to prevent further electrons and holes from moving across the junction, and a built-in voltage is created. Remember that the energy of these quantum states was found by the Schrödinger equation and is determined by the electrical attractions between the positively charged nucleus of each atom in the semiconductor and the negatively charged electrons. The effect of having an electric field across the interface between the n-type and p-type semiconductors is to raise the energy of the seats on the p-type side, relative to the

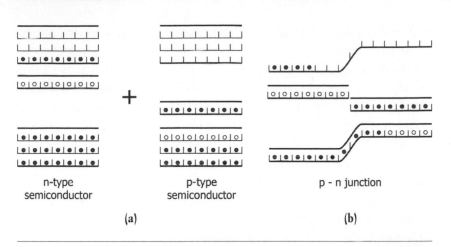

n-type   p-type   p - n junction
semiconductor   semiconductor

(a)   (b)

Figure 42: *Sketch of an n-type doped semiconductor and a p-type semiconductor (a) brought into electrical contact, enabling electrons from the n-type side to fall into holes from the p-type side, leaving behind positively charged mezzanine seats and negatively charged lounge seats in the n-type and p-type semiconductors, respectively. These charged mezzanine and lounge seats create a built-in electric field that affects the flow of electrical current through the semiconductor. The influence of this electric field is to tilt the two auditoriums, relative to each other (b). For simplicity only the first rows of the balcony and orchestra are shown in the figure to the right (b). If an external voltage is applied across this junction, it can cancel out this built-in field, making it easy for a current to pass from one side to the other.*

energy of the seats on the n-type side, as shown in Figure 42b. Electrons on the left will now find it harder to move over to the right, and holes on the right will find a barrier inhibiting them from moving to the left. I have now made one of the most revolutionary devices in solid-state physics—the diode.

The built-in voltage across the junction between the p-type and n-type semiconductors serves as a one-way door, like a turnstile that rotates in only one direction, for electrons.* Unidirectional valves are of course quite common, from the valves in your heart to the "cat's-whisker" rectifiers employed in early radio sets (used to convert an alternating current into a direct current). Solid-state semiconductor diodes are durable and small and can easily be tailored for specific electronic needs. If an external voltage is applied

---

* The arguments for electrons in the balcony in one direction will hold for holes in the orchestra in the opposite direction. For simplicity, I focus on the electrons.

across the junction with a polarity that is opposite to that of the built-in electric field, it will cancel out the internal energy barrier at the interface. The seats on the left- and right-hand sides will line up as if there is no built-in electric field. It will then be easy for electrons to move from the n-type side to the p-type side, and we will see a large current. If the direction of the voltage is reversed, the applied voltage adds to the built-in voltage, there will be a larger electric field opposing current flow, and the seats on the right-hand side are pushed up to an even higher energy. The diode then acts as a very high-resistance device. This directionality is important in radio detection or for power supplies, where an alternating-current input must be converted into a direct current.

One way to moderate the current passing through a diode is to vary the barrier that inhibits charges from moving from one region to another. In addition to barriers created by internal electric fields at the p-n interface, one can construct a diode where the two regions of an identical semiconductor are separated by a very thin insulator. The electrons cannot jump over this insulating partition, and the only way they can move across the interface is to quantum mechanically tunnel! Changing the voltage applied to the insulator has the effect of changing the height of the barrier seen by the electrons, and the tunneling current is a very sensitive function of the barrier height. In this way a small applied voltage can have a large influence on the flow of electrical current, and these tunneling diodes are an integral part of many consumer electronic devices, such as cell phones. Though I cannot predict whether any particular electron will pass through the barrier or not, when dealing with a large number of electrons I can accurately determine the fraction that will make it across. In this way electronic devices that rely on one of the most fantastic of quantum mechanical phenomena can be designed and counted on to operate in a routine, dependable manner.

*   *   *

A one-way door for electrical current is sometimes referred to as a "rectifier" and is useful for converting a current that alternates direction into one that moves in only a single direction, as in a radio receiver. In 1939 Russell Ohl, a scientist at Bell Labs, was studying the electrical properties of semiconductors for use as rectifiers when he examined a sample that accidentally contained a p-n junction. As he investigated the unusual current-carrying properties of this sample, he was surprised to find that a large voltage spontaneously appeared across the material when he illuminated it with a forty-Watt desk lamp. Ohl was looking for a rectifier, and he found a solar cell.

If I shine light on the semiconductor pn junction, as shown in Figure 43a, then as the energy separation between the orchestra and the balcony is the same on either side (the built-in electric field has just shifted the energy of the seats on one side relative to the other) photons will be absorbed on both sides, creating electrons in the balcony and holes in the orchestra. As everything we say about electrons in a diode will hold for the holes, but in reverse, we will focus on just the electrons. From the point of view of the electrons, if they are on the p-type side the internal, built-in electric field, it is as if they start at the top of an energy hill. These electrons will easily move down the hill in the balcony over to the left-hand side, and through the device. The electrons and holes are generated throughout the material, but those in the vicinity of the interface between the p-type and n-type semiconductors will see an internal voltage that will push the electrons down the hill and through the device. We will be able to draw a current, and obtain usable electrical energy, simply by shining light on the diode. So a solar cell is an illuminated p-n junction that generates a current.

And a light-emitting diode (LED) is a pn junction with a current that generates light. How do we get light from a diode? Remember that very far from the interface, the device looks like either a normal n-type semiconductor or a normal p-type semiconductor. In an n-type material there are a lot of electrons but very few holes (because the electrons were not excited from the filled band but were introduced by the chemical impurities). Thus, there is very little light generated when electrons in the nearly empty balcony fall to the orchestra, simply because there are very few vacant seats for them to fall into. If I force the electrons, by way of an external voltage, to move from the n-type side to the p-type side, there will be a lot of electrons in the middle junction region, where there will also be a lot of holes heading in the opposite direction

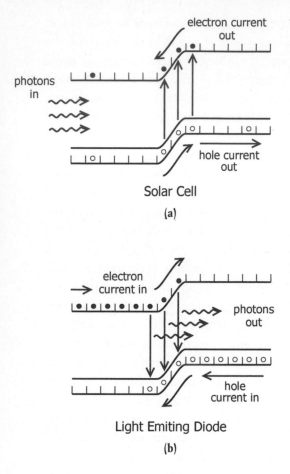

electron current out

photons in

hole current out

Solar Cell

(a)

electron current in

photons out

hole current in

Light Emitting Diode

(b)

Figure 43: *Sketch of a p-n-junction that absorbs light and promotes electrons up into the balcony, leaving holes in the orchestra. The electrons on the p-type (right-hand) side in the balcony are at a higher energy and can easily move down to the balcony states on the left-hand side of the material. An electrical current thus results from the absorption of light (a), and a diode in this situation is called a solar cell. Alternatively, if I force an electrical current through the diode, pushing electrons from the n-type side to the p-type side (and holes in the opposite direction), then at the interface region where the density of electrons and holes is roughly equal, there will be many opportunities for the electrons to fall from the balcony to the orchestra, emitting light as they do so (b). The device run in this way is called a light-emitting diode.*

(Figure 43b). As the electrons fall into the holes, they emit photons. It takes electrical energy to move the electrons up the hill into the p-type region, and we get some of that energy back in the form of photons. With a cylindrical transparent plastic lens covering the LED, we now have a light source without a filament (so nothing to burn out) and that is highly efficient (as very little electrical energy is converted to waste heat).

LEDs back in the 1960s could produce only red light at very low intensity. In the past thirty years the luminosity of these devices has increased by a factor of ten thousand, and LEDs that generate white light can put out as much illumination as an incandescent 60–watt lightbulb. LEDs are much longer-lived than standard incandescent bulbs, and some LEDs manufactured in the

1970s and 1980s are still operational today. There is interest in replacing silicon in these devices with organic chemicals as the semiconductor material since these organic light-emitting diodes (OLEDs) can be spread easily over large areas and in some cases are brighter than silicon-based LEDs. These devices use polymers consisting of long chains of carbon atoms bonded in a line, with different chemical groups protruding from the chain (referred to by chemists as "organic molecules") as the semiconductor element in the LED. There may come a day, and it may be relatively soon, when conventional incandescent lightbulbs are replaced with white-light LEDs, which are more environmentally friendly, use less power, and last much longer than compact fluorescent bulbs. All thanks to quantum mechanics.

If you'll allow me to digress, Fearless Reader, I'd like to relate a personal experience with diodes. When I was a new graduate student, I had a desk in the research group where I would eventually conduct my dissertation studies. A more senior student, Peter, had recently convinced our research adviser to install a separate phone line in the lab where Peter worked. This was back in the days before AT&T had been broken up into the "Baby Bells" by court order, and one had to order phones from Bell Telephone, which would then come and install the landline (cell phones were an extreme rarity back then). As a cost-saving measure, our professor had ordered a phone that would allow incoming calls but not outgoing ones. As soon as the phone company employees departed, Peter took the phone off the wall, opened it up, and compared it carefully to another phone that did allow outgoing calls. At one key juncture Peter noted—a diode! A simple and elegant way to ensure that signals could propagate in one direction but not in the other. Removing this semiconductor device and reconnecting the wires, Peter was then able to make outgoing calls on the phone.

All of which I was unaware of when the next day, a Saturday, the phone company employees returned to check on the line and attach the sticker to the phone that would indicate its assigned phone number. A short while later the phone in the common room where I was studying rang, and when I answered it, the telephone workers apologized, saying that they just wanted to make sure that the phone could make outgoing calls. A short while after they left,

Peter came in and told me about his "fixing" of the phone in his lab. He turned white when I mentioned that the phone company men had been by earlier, wanting to make sure the phone could make as well as receive calls. Racing to his lab, he quickly ascertained that the phone would no longer make calls. I watched as he took the phone off the wall, removed the plastic cover, and spotted a small white card *inside* the phone, tucked underneath a tangle of wires. Removing the card with tweezers, we saw the handwritten suggestion: "Try it now, whiz kid."

Sure enough, the phone had been completely rewired, with great effort going into replicating the functions performed by one simple diode. Without the innovation of the solid-state diode, electronics would be a great deal more complicated and bulkier—though the phone company would still always win in the end!

CHAPTER SEVENTEEN

Big Changes Come in Small Packages

The diode was the first important step into the semiconductor age, and the second major advance also came from Bell Labs, with the invention of the transistor. There are two different structures for a transistor—one that involves adding another n-type semiconductor to the pn junction (to make an npn device) and the simpler-to-understand (in my opinion) field-effect device. I discuss here the field-effect structure, as it will also help us understand how a flash memory stick (also known as a jump drive or a USB drive) works.

The arguments here do not depend on whether the material is n-type or p-type, but for simplicity we will pick a p-type semiconductor for definiteness. Imagine a semiconductor, such as silicon, as a rectangular slab, longer and wider than it is thick, as illustrated in Figure 44a. On part of the top of the semiconductor we place an insulator that could be silicon dioxide, which in its crystalline form is called "quartz" and in an amorphous phase is termed "glass." On the top of the semiconductor surface there are two metal electrodes at each end, not touching the insulator. At one electrode a voltage is applied, and the resulting current pushed through the semiconductor is withdrawn at the other electrode. So far we have just described a way to measure the current passing through the material for a given applied voltage, and the insulator plays no role in the conduction through the semiconductor. The metal electrodes on either side of the insulator have a surfeit of free electrons, and where they are in electrical contact with the p-type semiconductor they form an effective p-n junction. If we are trying to flow electrons through a p-type doped material these back-to-back p-n junctions will make it very difficult for the current to move through the semiconductor. Now, to make this device a transistor, let's put a sheet of metal atop the insulating slab. As shown in Figure 44a, there are two metal electrodes apart from each other on the top of the semiconductor, between which is an insulating slab, atop of which is another metal electrode. We have constructed a field-effect transistor.*

What happens if we apply a positive voltage to the metal electrode that is covering the insulating slab? As this top metal electrode is in contact with an insulator that does not conduct electricity, the charges will just stay on the metal, having no place to go. The electric field created by these charges will extend through the insulator into the semiconductor layer. Compared to the insulator, the semiconductor beneath the insulating slab is a pretty good conductor, though not as good as the metal. Electrons will be drawn toward the region near the insulator-semiconductor interface by the electric field. Without a voltage, the metal is electrically neutral, and there is no reason for any electrons in the semiconductor to be pulled toward this region. With a voltage applied to the insulator, piling up positive charges on the top of the insulating slab, a channel of electrons connecting the two other metal electrodes at either end of the semiconductor is created, shown in Figure 44b. This has the effect of reducing the electric field at each pn junction where the metal electrodes contact the semiconductor, and the ability of the material to carry a flow of electrons will be greatly improved. The positive voltage on the metal atop the insulator in a sense opens a gate through which the electrons can flow. Applying a negative voltage would push electrons away from the insulator-semiconductor interface, and the ability for electrons to flow would be reduced (the gate would swing shut in this case).

---

* There are many different types of transistor structures—what we have described is technically referred to as a metal-oxide-semiconductor field-effect transistor, or a MOSFET.

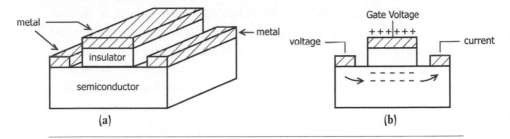

Figure 44: Sketch of a simple transistor device structure (a). Two metal electrodes on the top of the semiconductor are used to pass a current through the device. A thin insulator (such as glass), on which is a metal electrode, lies on top of the semiconductor between the two metal electrodes used to pass the current. When a positive voltage is applied to the "gate electrode," positive charges accumulate on the top of the insulator, which attract electrons in the semiconductor to the region underneath the insulator (b). These electrons improve the ability of the semiconductor to pass a current between the two metal electrodes, and the current is made much larger by the application of the "gate voltage."

This is what a transistor does—it provides a way, by applying a small voltage to the gate electrode, to dramatically alter and potentially amplify a current passing through the semiconductor.

Remember from the last chapter the discussion of the influence of a built-in electric field on a pn junction on the energy of the bands of states in a semiconductor. Changing the electric field alters the energy of the quantum states that are calculated using the Schrödinger equation (demonstrated through the influence of the electric field of the positively charged nucleus on the electrons' allowed energies). If no voltage is applied, then there is no extra field on the semiconductor, and the number of electrons available to flow in the p-type semiconductor is very low. If a positive voltage is applied to the insulator, it will change the energy of both the orchestra and balcony of states. The change is strongest near the positive charges on the insulator and decreases as one goes farther into the semiconductor. Near the region by the insulator, electrons can now be thermally promoted into the balcony of the material. Consequently, the region near the insulator, when there is a positive voltage applied to a p-type semiconductor, will see a large enhancement in its ability to carry a current of electrons. The sensitivity of the current passing through this device to an externally applied gate voltage results from changes in the energy of the quantum states in the filled orchestra and empty balcony, which in turn are understood from the quantum theory of solids.

If the voltage applied to the insulator changes with time (such as in the case of a weak radio signal detected by an antenna), then the current passing through the semiconductor will also vary in time but as an amplified version, capable of driving speakers so that the radio signal can be heard. Transistor radios and television sets, employing the amplification capability of these devices, were some of the first applications of these devices, replacing the vacuum tubes and making these consumer electronic products smaller and lighter.

Vacuum tubes accomplish the same task as a transistor, by heating a wire until electrons "boil off" the filament. A voltage applied to a screen then attracts these free electrons to a collector, and depending on the voltage, the electrons can be accelerated toward or reflected away from the collector. In order to minimize the electron beam's scattering from air molecules, all the air in the tube should be removed. Such devices are bulky and fragile, use considerable power, generate a great deal of heat (necessitating spacing them a distance from each other), and are expensive to produce. A semiconductor transistor accomplishes the same task without requiring a glass-enclosed vacuum, in a compact, rugged design, and wastes very little energy as heat; and the only limitations on the size of the device are the ingenuity in constructing the insulating slab and applying the metal electrodes, and making contact with the rest of the circuit.

If we can make the transistors small, we can put several transistors on a single piece of silicon. By varying the concentration of chemical impurities that add either excess electrons or holes to the semiconductor, and through the placement of other metal electrodes and insulating slabs, one can incorporate diodes, resistors, and capacitors into the same semiconductor along with the transistor. In this way the various aspects of a complex circuit can be integrated onto a single semiconductor chip. In 1958, just a year after the Challengers of the Unknown faced off against ULTIVAC, whose electronic brain was as large as a room (Figure 39), Robert Noyce and

Jack Kilby independently designed and constructed the first integrated circuits. The first of these devices incorporated roughly five
to ten transistors on a single silicon wafer. In the introduction I
discussed Moore's law, whereby the number of transistors on an
integrated circuit doubles every two years. The continued accuracy
of this prediction surprised even Moore, and in 2010 the number of
transistors on a chip can be over a billion. Estimates of the number
of transistors in a computer's microprocessor suggest that on a
typical college campus there are many more transistors than there
are stars in the Milky Way.

These transistors do more than simply amplify information, as
in the case of a weak electromagnetic wave signal being boosted
in a cell phone. They also can store and manipulate information.
When a large positive voltage is applied to the gate electrode on the
insulator, the current-carrying capability of the semiconductor is
greatly enhanced. Removing this large voltage restores the silicon
to its poorly conducting state. The first situation can be described
as a "one," while the second is a "zero." Just as the DVDs and CDs
in the previous chapter are able to encode complex information
through a series of ones and zeros, the transistors on the integrated
circuits can do the same. However, transistors offer the possibility of much greater sensitivity to small perturbations. Transistors
have been fabricated in the research laboratory with dimensions of
under a hundred nanometers, where, depending on the voltage applied to the gate electrode, the transport of a single electron can be
detected. Computers use transistors as logic elements that can be
in an "on" or "off" state—and transistors can be fabricated whereby
the difference between the two conditions is the motion of one
electron.

It may seem surprising that one could construct a complex literature using an alphabet consisting of only two letters. But if there
is no constraint on the length of a given word, then there is indeed
enough flexibility to perform even the most sophisticated mathematical operations, such as adding two numbers. A full discussion
of how diodes and transistors are combined to perform a variety of
logic functions, and the Boolean mathematics that underlies their
calculations, warrants its own book, and would take us too far
afield for a discussion of the applications of quantum mechanics.

Nevertheless, I do want to conclude this chapter with a discussion of one modification of the transistor structure that is already changing our everyday life.

Long-term information storage in a computer is done via the magnetic hard drive. A disc contains a record of "ones" and "zeros" in the form of magnetic domains, with a magnetic field pointing in one direction counting as a "one" and a field pointing in the opposite direction as a "zero."* An externally applied magnetic field can polarize regions ("bits") on the drive, and write the sequence of ones and zeros that encode information. A smaller magnetic sensor, essentially a layered metallic structure whose resistance is very sensitive to the external magnetic field, is brought near the disc. If the magnetic field of a bit points in one direction, the resistance of the sensor will have one value; it will have another if the bit's magnetic field is in the opposite direction. The disc itself spins like a DVD or CD at high speeds of more than five thousand revolutions per minute, and the sensor rides just above the hard drive, with a spacing that is equivalent to one-hundredth the diameter of a human hair. To store more information, one makes the platter larger and the bits smaller. That this magnetic device is capable of storing information without an external power supply (once the bits are magnetized, they stay in the same orientation), and with a relatively low failure rate (despite fears of "hard-drive crashes," the medium is extremely reliable given the use it endures), is a testament to the skill of engineers.

Transistors are also able to store information. For the device configuration we have discussed, the conductance of the semiconductor is low if there is no voltage to the gate metal (representing a "zero"), and high (standing in for a "one") when a positive voltage is applied. However, once the external voltage is turned off, then all transistors in a circuit default back to their low conductance state.

How can I store and preserve the high-conductance channel of a transistor, that is, keep the "ones" from turning into "zeros"

---

* For technical reasons the actual bits on a hard drive involve several magnetic domains, arranged in different sequences to represent "ones" and "zeros," but for our purposes we can simplify this to single domains for a "one" or a "zero."

after the voltage is turned off? Flash memory devices add a very small wrinkle on the field-effect structure we have described. To the standard field-effect transistor configuration, the flash memory adds a second metal electrode in the insulating layer, a very small distance above the semiconductor. So the device has a metal gate, a thin layer of insulator, another thin metal electrode, and then a very thin layer of the insulator atop the semiconductor.

What's the point of the second metal layer? If the two electrodes that used to pass the current through the semiconductor are shorted, and a large voltage is applied to the gate metal, then electrical charges can quantum mechanically tunnel to this interior electrode. This electrode is not connected to any outside wires and is termed the "floating gate." The floating gate can be a thin metal film, or it could be a layer of silicon nanocrystals, separated from one another so that these charges remain on the silicon particles and do not leak away. The charged floating gate generates an electric field in the semiconductor, influencing the current-carrying channel and maintaining the device in either a high- or low-conductance state (that is, recorded as a "one" or a "zero") even after the voltage is removed from the gate metal. Until a voltage of opposite polarity is applied, the transistor will store this state of the transistor, even when the transistor is unplugged from any power supply (such a memory is termed "nonvolatile"). The story goes that a colleague of Fujio Masuoka, the inventor of this type of transistor memory, when describing how quickly the stored information could be erased, said that it reminded him of a camera's flash, whence the nickname for the device derives. At the time of this writing, flash memory devices capable of storing 256 gigabytes of information (large enough to store more than ten thousand copies of this book as Word documents) are being manufactured.

Nonvolatile memories have also revolutionized photography. In conventional, nondigital cameras, a light photon induces a chemical change in a photographic film. The information as to where the photon was absorbed by the molecule in the film is stored, and then a series of wet chemistry steps transfers this information to a photographic print. The graininess of individual molecules in a conventional film is now replaced with a pixilated grid. When photons strike the photodetectors in a given pixel, they will, if absorbed, create mobile charges. Using different semiconductors, the energy separation between the filled and empty bands of states can be changed, enabling photodetectors that can image in the infrared, visible, or ultraviolet portion of the spectrum. The charges up in the balcony can be converted to voltages, and then stored on flash memories. The location of each pixel is known, so a digital record of the number of photons that struck the array of photodetectors is obtained.

Once an image is digitally captured, the ability to display it on a flat panel screen, as opposed to the bulky cathode ray tubes that were a feature of televisions up until fairly recently, also makes use of semiconductor transistor technology. The bits of information in this case are the pixels on the display screen. In each pixel is a small amount of a "liquid crystal," consisting of long chain organic molecules (that is, carbon atoms bonded in a line, with various other elements and chemical groups protruding from the carbon chain). Geometric constraints and electrostatic charges along the carbon line will lead certain long chain molecules to pack together in different arrangements, from a loose, random collection to a herringbone pattern not unlike a professor's tweed coat to a more ordered phase similar to matches tightly stacked in a box. Just as the matches can be easily poured out of the matchbox regardless of their packing, these long chain molecules retain the ability to fill a container and flow as a fluid.

Certain liquid crystal molecules will make a transition from one ordered configuration to another when the temperature is changed—or if an external electric field is applied to the molecules. The optical properties of water change dramatically when ice undergoes a phase transition and melts—similarly, when certain liquid crystals change from one packing state to another under an external voltage, there can be an associated change in their optical properties, such as whether the material reflects light and is shiny or absorbs light and appears dark. Early "liquid crystal watches" had metal electrodes in "broken eight" pattern, and depending on which metal plate had an applied voltage, different regions of the liquid crystal would appear dark, and thus form different numerals depending on the time of day. These liquid crystal displays (LCDs) are still employed in certain clocks and timers. For more sophisticated image displays, a capacitor and a thin film transistor (sometimes referred to as a TFT) are placed behind each liquid crystal pixel. Color filters can convert a grayscale image to a color one, and by changing the timing of when each pixel is turned on and off, one can view a moving image, similar to the television screen shown on the cover of the December 1936 *Amazing Stories* science fiction pulp (seen in Figure 45).

The ability to instantly display the stored image (or video) and the convenience of data transfer and large storage capacity, coupled with the incorporation of these cameras into other devices (such as cell phones or computer screens), has exceeded the expectations of science fiction pulp magazines—well, with one exception. As illustrated in Figure 46, the notion that a device capable of wireless video reception and broadcasting small enough that it would fit on a person's wrist was indeed anticipated in 1964 by the comic strip creator Chester Gould. Wrist phones that are capable of video

© 1936 Teck Publications Inc.

Figure 45: *While the Space Marines appear to be viewing a flat panel display on this cover, the story by Bob Olson indicates that they are in fact watching a 3-D picture tube image.*

© 1964 Tribune Media Services, Inc.

Figure 46: *Dick Tracy using a two-way wrist phone with video capabilities, This gadget was introduced in 1964, a good forty years before real technology would catch up with the comic strips.*

transmission are now becoming available, another example of fiction becoming reality through quantum mechanics. Now, if we could only figure out how to construct personal "garbage cans" (Chapter 4, Figure 8) that fly by means of magnetism!

CHAPTER EIGHTEEN

Spintronics

Everything—light and matter—has an
"intrinsic angular momentum," or "spin,"
that can have only discrete values

One of the most surprising discoveries made by physicists probing the inner workings of the atom was that electrons—subatomic particles that are the basic carriers of negative charge—also are little bar magnets, like those shown in Figure 10 in chapter 4. This intrinsic magnetic field is associated with a property called "spin," though this term is a misnomer—while it does relate to intrinsic angular momentum, the magnetic field associated with the electron doesn't really come from its spinning like a top. Nevertheless, when physicists refer to the internal magnetic field possessed by electrons (or protons or neutrons), they inevitably speak of the particle's spin.

A transistor modulates the current flowing through a semiconductor by the application of an electric field to an insulating slab on top of the conducting material. In this way the current flowing through the semiconductor is regulated through the electron's negative charge. The magnetic field that the electron exhibits has been, in most electronics up till now, completely ignored. As one might imagine, this situation changes in devices characterized as "spintronic," a shorthand expression for "spin transport electronics." Here the electron's magnetic field is a crucial component of the signal being detected or manipulated. One form of spintronics has been employed in computer hard drives, while the next generation of such devices (discussed in Section 6) may make hard drives unnecessary.

As described in Chapter 15, a DVD encodes information in the form of ones and zeros as smooth or pitted regions on a shiny disc. A laser reflected from the surface of the disc does so either specularly, that is, smoothly onto a photodetector if the surface is smooth, or diffusely, away from the detector if it strikes a jagged pit. Similarly, the hard disc drive in a computer is a magnetic material with regions magnetized in particular patterns; the smallest elements of the pattern are termed "bits." The drive stores information in the form of ones and zeros as magnetized regions, with north poles pointing in one orientation representing a "one," and in the other direction standing for a "zero." Each bit (in current disc drives) is written by moving a magnet over the region, which orients the magnetic pattern. To create the opposite pattern, a magnetic field in the reverse direction is applied. To erase the bit, a depolarizing magnetic field is applied. To read the "one" or "zero" stored on the disc, hard drives employ sensors such as "giant magnetoresistance" devices or "magnetic tunnel junctions."

All solids have bands of allowed states in which the electrons may reside, separated by energy gaps where there are no allowed quantum states. The difference between an insulator and a metal is that for an insulator (or a semiconductor), the last filled band, the orchestra in our auditorium analogy, is completely filled, with every possible energy state being occupied by an electron. In contrast, in metals, the lower orchestra level is only half filled, as shown in Figure 34b in Chapter 14. If a voltage is applied to a metal, the electrons feel a force. This force in turn accelerates the electrons, causing them to speed up and increase their kinetic energy. Recall the water-hose analogy of metal wires—the voltage is like the water pressure, and the electrical current is the resulting flow, of water through the hose. As there are always some unoccupied seats in the lower orchestra level of a metal, electrons in the upper, filled rows can always move to higher energy states, and the material is able to conduct an electrical current.

What determines the current observed in a metal for a given applied voltage? Normally the electrons can surf using the atoms in the metal wire—as long as the atoms are in a uniform crystalline arrangement, they do not impede the electrical current. One can run on a city sidewalk and never step on a crack (thereby preserving one's mother's back) as long as the placement of the concrete segments is uniform and matched to one's stride. If there is a hole in the sidewalk, or a protruding tree root, or a shortened segment, then it is likely that the runner will stumble. In any real metal wire there will be defects such as crystalline imperfections (atoms randomly located out of their preferred ordered positions) and impurities that inevitably sneak into the solid during the fabrication process. Electrons accelerated by a voltage will scatter from these defects and transfer some of their kinetic energy to these atoms.

Sometimes this scattering is a good thing, as in an incandescent lightbulb or a toaster. There a large current is forced through a narrow filament, and the accelerated electrons transfer so much of their energy to the atoms in the wire that they shake violently about their normal crystalline positions. This shaking heats the wire until it is glowing red-hot (as in the coils in your toaster), and for higher currents in thinner wires, the shaking can cause excitation of electrons to all higher energy states equally, with resultant emission of light of all frequencies, perceived as white light (as in the filament of a lightbulb). Sometimes the loss of energy through collisions with atoms in the metallic wire is a bad thing, as in electrical power transmission cables; in order to compensate for these energy losses, the voltages along the lines must be very high, requiring power substations and transformers along the line.

Computer hard-drive disc readers employ the scattering of an electrical current by magnetic atoms to sense the different fields of the magnetized bits. A thin, nonmagnetic metal is sandwiched between two magnetic metals. In the absence of an external magnetic field, one slice of magnetic "bread" is permanently polarized so that its magnetic field points in one direction within the layer, while the other slice of bread is polarized in another direction (the nature of the quantum mechanical coupling between the magnetic layers, separated by the nonmagnetic middle layer, leads to this configuration being the low-energy state).

Imagine a flow of electrons perpendicular through the top of this "sandwich," passing through the face of one slice of magnetic bread, through the nonmagnetic metal meat of the sandwich, and finally through the face of the other magnetic metal bread slice, as shown in Figure 47. When first entering the first magnetized layer, the electrons are unpolarized—their internal magnetic fields are as likely to point in one direction (spin "up") as the other (spin "down"). The first ferromagnetic layer polarizes the electrons, and those that move into the nonmagnetic spacer layer will have their internal magnetic fields pointing in the same direction as the field in the first metal layer. When they reach the second magnetized layer, which normally has a field pointing in the opposite direction, these polarized electrons are mostly reflected, so very little electrical current passes through the second layer and leaves the sandwich. If very little current results for a given voltage, we say that the device has a high resistance for an electrical current passing perpendicular through the three layers.

Now this structure is placed in an external magnetic field, such as that created by a magnetized bit on a computer hard drive. The external field forces both magnetic layers in this sandwich

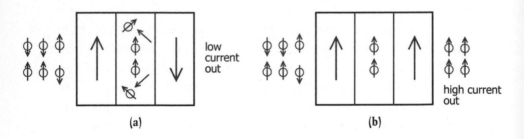

(a)   (b)

Figure 47: *Sketch of the device structure used to measure magnetic fields with an electrical current in a computer hard drive. An electrical current has both a negative charge and a built-in magnetic field resulting from its quantum mechanical intrinsic angular momentum ("spin"). Electrons flowing into the device are magnetically polarized by the first layer. In (a), the second layer is aligned opposite to the first, so the electrons polarized by the first layer are repelled by the second, and a very small current results. In the second case (b), the second magnetic layer is aligned in the same direction as the first, and the polarized electrons easily pass through the second layer. This configuration would present a low resistance to the flow of current, while the first case (a) would represent a high resistance state.*

(Figure 47b) to point in the same direction. When an electrical current now passes through this structure, the first layer polarizes the electron's magnetic fields as before, and the second layer, now pointing in the same direction, readily allows the electrons to pass through, and hence a large current flows through the three-layer device. This change in resistance with an external magnetic field can be very large, up to 80 percent or more (they are, seriously, technically known as giant magnetoresistance devices), which means that they are very sensitive to even small magnetic fields. One can thus make the magnetically polarized bits on the hard drive smaller and still be able to reliably read out the sequence of "ones" and "zeros." Smaller bits means more of them can be packed on a given disc area, and the storage capabilities of computer hard drives have increased dramatically since the introduction of this first spintronic device.

The first generation of iPods was able to store large data files on a small magnetic disc because the sensors used to read the information made use of the giant magnetoresistance effect. The drive to pack smaller magnetic bits at higher densities has led to the development of magnetic sensors on hard drives that employ another quantum mechanical phenomenon—tunneling—to sense the magnetic fields of the bits. These sensors have essentially the same structure as the device in Figure 47. Instead of a nonmagnetic metal placed between the two magnetic slices of bread, a thin insulator is used. A current can pass through the device only via tunneling, and the probability of this process turns out to be very sensitive to the magnetic polarization on either side of the barrier. These devices provide an even more sensitive probe of very small magnetic fields and are found in computer hard drives currently available for purchase. Every time we access information on our computers, we are making use of the practical applications of quantum mechanical tunneling.

The basic principles underlying giant magnetoresistance are finding new applications in future spintronic devices. Giant magnetoresistance was discovered in 1988 by Albert Fert in France and independently by Peter Grünberg in Germany, for which they shared the Nobel Prize in Physics in 2007. By 1997, hard drives containing read heads using the giant magnetoresistance effect were available for sale. It is actually not unusual for quantum-mechanics-enabled devices to quickly find their way into consumer products. Bell Labs held a press conference announcing the invention of the transistor in 1948, and by 1954 one could purchase the first (expensive) transistor radio.

CHAPTER NINETEEN

A Window on Inner Space

In the 1963 Roger Corman science fiction film *X: The Man with the X-ray Eyes*, Dr. James Xavier, searching for improvements in patient care, develops a serum in the form of eye drops that enables a person to see through solid matter. Eschewing animal testing as not being suitably reliable, he experiments on himself and does indeed gain the ability to see through a person's clothing and epidermis. However, this success leads to one of the greatest catastrophes that can befall any scientist—he loses his research grant when his funding agency discounts his claims of "X-ray vision!" Nevertheless, his ability to see within the interior of a person enables him to save a small child's life, as he recognizes that she was about to receive an unnecessary and ineffective operation. Sadly for Dr. Xavier, his X-ray vision becomes stronger and stronger, until his eyelids and thick dark glasses provide no respite. It does not end well for the well-meaning doctor, as the biblical expression "If thine eye offend thee . . ." plays a key role in the film's conclusion.

Fortunately we can safely peer inside a person, see his or her internal organs, and discriminate healthy tissue from cancerous growths, without the disastrous consequences suffered by Dr. Xavier. I now address a device that has become common in most hospitals and many medical clinics and would certainly have strained the credulity of the editors of any science fiction pulp magazine had it been featured in a submitted story—magnetic resonance imaging, or MRI. This process, enabling detailed high-resolution imaging of the interior of a person, is a striking illustration of how our understanding of the quantum nature of matter, driven by scientists' curiosity in the 1920s and 1930s about the rules governing the behavior of atoms and light, has led to the development of technologies that futurists could not suspect fifty years ago.

We have made use of the intrinsic angular momentum of fundamental subatomic particles when determining which form of quantum statistics—Fermi-Dirac or Bose-Einstein—they would obey. In this way the internal spin is crucial to understanding the nature of metallic or insulating solids but was not employed directly when describing the physics of diodes or transistors. Associated with the spin is a small intrinsic magnetic field that enabled Stern and Gerlach to measure the spin in the first place (Chapter 4), and remotely probing the magnetic field from the spin of nuclear protons enables magnetic imaging.

We are composed mostly of water—molecules consisting of an oxygen atom bound to two hydrogen atoms. Each hydrogen atom has a single proton in its nucleus. The intrinsic spin of the proton is $\pm\hbar/2$, and there is a small magnetic field associated with each proton. The magnetic field has a north and south pole (Chapter 4, Figure 10), and when we place the water molecule in an external magnetic field, the hydrogen atom's proton points either in the same direction as the applied field (that is, its north pole points up while the lab magnetic field does likewise) or in the opposite direction (its north pole points down while the external field's north pole points up). The proton in the nucleus has a series of available quantum levels, just as the electron has its own series of possible states. If the proton's magnetic field is in the same direction as the external magnetic field, it is in a lower energy state. If the proton's magnet is opposite to the external magnetic field, then it will have a higher energy, as it takes work to rotate it to the lower-energy, aligned configuration. As indicated in Figure 48, the energy level that is occupied by the single proton can thus be split into two energy values by placing the hydrogen atom between the poles of a strong magnet, with the proton's energy being lower than what we find with no outside field or the energy being higher, depending

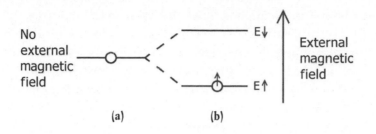

Figure 48: *Sketch of the energy level of a single proton in the nucleus of a hydrogen atom (a) when no outside magnetic field is applied and (b) when a field is present. In the second case the proton's energy is lowered if its own intrinsic magnetic field points in the same direction as the outside magnet, and the energy is higher if it points in an opposite direction. In this figure the proton is indicated with its spin aligned with the external magnetic field and thus in the lower energy state. If the spin were opposite to the external field, the proton would reside in the higher energy state.*

on whether the proton's magnetic field points with or against the external field, respectively.

Say a proton is aligned in the same direction as the external magnetic field, in the lower-energy split state (Figure 48b). If I provide energy, in the form of a photon, I can promote the proton to the higher-energy state, which corresponds to the proton's magnetic field being opposite to the outside magnet. This resonant absorption is entirely analogous to the line spectra (Chapter 5, Figure 13) for electronic energy levels in an atom. In essence the photon provides energy to flip the proton's internal magnet, from pointing up to pointing down (for example). It is as if I had a top that was spinning clockwise, and with an appropriate pulse of energy I caused its direction to reverse, so that it was now rotating counterclockwise.

The bigger the external magnetic field, the larger the energy splitting. That is, it requires more energy to flip the orientation of the proton's magnetic field if it is in a large external field than in a weak field. If the hydrogen atom is placed in a magnetic field roughly twenty thousand to sixty thousand times stronger than the Earth's magnetic field,* then the separation in energy between

---

* Which is why one must avoid stray metallic objects in the MRI room when the magnet is powered.

when the proton is aligned with and against the external field is less than a millionth of an electron Volt. In comparison, the binding energy holding the hydrogen atom to the oxygen atom in the water molecule is nearly five electron Volts. Recall from Chapter 2 Einstein's suggestion that the energy of a photon is proportional to its frequency ($E = h \times f$). A photon capable of promoting the proton from one magnetic orientation to the other (as in Figure 48b) is in the radio portion of the electromagnetic spectrum. As this form of electromagnetic radiation penetrates through a person (which is why you can hear your transistor radio even when you place your body between it and the broadcast antenna), this energy region is well suited to probing the proton's orientation within a person.

The idea begins to form. Place a person in a large magnetic field, strong enough to generate an appreciable energy splitting for the protons in the water molecules that are in every cell in his or her body. Direct a transmitter of radio waves at the person, and the more photons that are absorbed, promoting a proton from one magnetic orientation to the other, the more water molecules there must be. How do we determine whether a radio wave is absorbed or not? A high exposure will transfer many protons from the lower energy state to the higher energy level. When the radio-frequency light is turned off, the hydrogen atom's protons flip back down to the lower energy state, emitting photons as they do. In this way the person "glows in the dark," emitting radio-frequency light that is detected and is the hallmark of the resonant absorption. The number of aligned protons throughout the bulk of the person's body may thus be measured.*

How do we obtain spatial resolution throughout a cross section of the person? By varying the strength of the magnetic field. Make the magnetic field very small at the left-hand side of the person and very large on the right, increasing linearly from one side to the other. As the energy spacing depends on the strength of the external magnet, the separation between levels will be small on the left and grow to a larger energy gap on the right. Consequently, the

---

* In fact, the radio-wave signal is turned on and off continuously, but the emission of radio waves when the system relaxes back to the lower-energy configuration is what is detected.

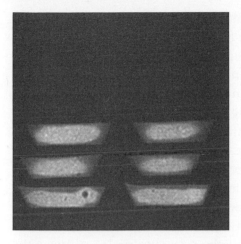

Figure 49: *Magnetic Resonance Image of three packets of peanut butter cup candy. The difference in spin relaxation times provides a basis for contrast between the chocolate coating and the interior filling, confirming the presence of the peanut butter inside the candy without having to bite into the candy (not that we wouldn't be willing to make such sacrifices for science!). Courtesy of Professor Bruce Hammer at the University of Minnesota.*

minimum photon energy that will induce a transition will be larger on the right than on the left. By varying the frequency of the radio signal, one can determine the amount of absorption on the left, middle, and right of the person. By using secondary magnets in the cylinder that encloses the person, information on the proton density with full spatial resolution can be obtained. As shown in Figure 49, this imaging of the magnetically induced resonant absorption enables us to probe the inner secrets of a three-pack of chocolate peanut butter cups, confirming, through application of advanced quantum mechanics, that there is indeed delicious peanut butter within the chocolate coating.

But all cells in the body contain water, so there will be strong proton absorption at all points in the body. Where does the contrast come from? There are other elements that exhibit a magnetically induced resonance absorption signal, such as sodium, phosphorus, carbon, oxygen, nitrogen, and calcium. These elements have radio resonances at different frequencies than hydrogen and can thus be distinguished from the single proton signal. However, a more powerful technique involves not the magnetically induced signal, but the manner in which it goes away and then returns.

When the magnetic field is applied, many of the hydrogen atom's protons in the water molecules will line up with the external field, so that nearly all of the lower energy states will be occupied, and the higher energy states, corresponding to the proton's magnetic field opposing the applied field, will be less occupied.* Under a continuous exposure of radio-frequency light, more and more protons are promoted to the higher energy state, until the situation is reached when the average number in the upper state (with a magnetic field pointing down) is equal to the number in the lower state (with a magnetic field pointing up). At this stage, we have for the collection of protons an equal number with their north poles pointing up as we have with their north poles pointing down. The total net magnetization of the protons will therefore be zero. If we now stop the continuous illumination, the protons in the higher energy state will relax back to the lower energy configuration. The characteristic time that this will take is highly sensitive to the local environment in which the particular water molecule resides, as the interaction of the proton's magnetic field with thermal vibrations of other atoms and with the magnetic field from other elements in its vicinity determines how hard or easy it is to polarize. It was discovered through careful experimentation that the time dependence of the restoration of the net magnetization is different for the various tissues in the body, providing a basis for contrast in the resulting images. One can inspect for blood-vessel blockages, cysts, or growths, and determine whether or not tumors are benign, based on differences in magnetization times. By careful examination of the time dependence of how the protons resume their original magnetization, a thorough diagnosis, which previously would have required X-ray eyes, is possible.

There are of course a host of complex technical issues that go into generating a three-dimensional image using MRI. I never promised I would tell you how to construct your very own imaging device, only that I would explain the essential quantum mechanics that underlies such a process. Needless to say, all of the above would be useless without high-speed computers utilizing solid-state integrated circuits, to record, store, and analyze the radio-frequency

---

* As the person is at a temperature of 98.6 degrees Fahrenheit, there is enough thermal energy to promote protons into the higher energy state, even without radio-frequency light. Increasing the external magnetic field raises the energy separation between states, and fewer protons will be found in the upper level in the dark. The detection sensitivity in an MRI device is quite high, and the MRI system is able to detect the small additional fraction of protons that are promoted by the radio waves.

absorption data. So in a sense, quantum mechanics enables MRI machines at two separate levels.

Dr. James Xavier could have saved himself a great deal of grief with his experimental eye drops by using an MRI to perform diagnoses on patients by peering at their inner organs. Similarly, Professor Charles Xavier (no relation), mutant leader of the X-Men and the world's most powerful telepath, can read people's thoughts. While this is not possible, by employing functional magnetic resonance imaging (fMRI), we can determine what regions of the brain a person is using, and from that make inferences as to what they are thinking.*

All cells in your body have a function and require energy when carrying out their designated tasks. Nerve cells—neurons—process information through the generation and transmission of voltages and ionic currents. When we eat, we ingest molecules originally generated by plants that contain stored chemical energy. The plants utilized the energy in photons from the sun to construct complex sugar molecules. The mitochondria in every cell synthesize adenosine triphosphate (ATP) from these sugars and thereby release some of that stored energy, which the cell can then use to perform various functions. The chemical trigger for the construction of ATP is the incorporation of oxygen molecules and the release of carbon dioxide. Consequently, whenever a cell is actively working, in particular for a prolonged period of time, there is an increase in blood flow to this cell, in order to maintain sufficient oxygen levels for ATP production. By looking where the blood is flowing, we can determine those cells that are most active.

Neurons do not store glucose, and consequently within a few seconds after you start some heavy thinking, there is an increase in blood flow to the region containing the active neurons. The brain has the same relationship to the body as the United States has to the rest of the world—the brain is roughly 3 percent of the total body mass, while it consumes 20 percent of the energy expended. You use different portions of your brain depending on the task you are performing—sitting while reading this book, walking while

---

* This technique is still in the experimental stage, so you don't need to invest in tinfoil hats just yet.

reading this book, or showering while reading this book. Thus, by monitoring which regions of the brain are receiving more oxygenated blood, one can ascertain what activity a person is engaged in, without having to look at the person directly. The true power of this technique for determining brain activity from the variations in blood flow involves distinguishing between cerebral tasks—mentally doing arithmetic compared to recalling a pleasant summer memory. In this situation the beatific smile on the person's face would not betray which task was being performed.

When an MRI scan is taken of a person's brain, spatial resolution on the order of millimeters and time resolution of one to four seconds are possible. When red blood cells are carrying oxygen, they are diamagnetic—which means that their internal magnets orient opposite to the direction of an external magnetic field. Conversely, deoxygenated hemoglobin is paramagnetic; that is, it will align in the same direction as an external field but will have no net magnetization in the absence of an applied magnetic field. Using magnetic resonance imaging, one can examine a region deep within the cerebral cortex and determine its rate of blood flow. By measuring the time dependence of the variation in magnetization when the saturating radio-frequency radiation is removed, one can distinguish which regions of the brain are in high need of additional oxygen and energy.

Alfred Bester wrote of the difficulty that Ben Reich faced plotting and carrying out an undetected murder in a society where nearly everyone, particularly the police, is telepathic in his 1953 novel *The Demolished Man*. The ability to read minds, to know what another person is silently thinking, has long been a hallmark of science fiction stories, predating the pulps and continuing into the present. The cooperative behavior exhibited by the characters in Theodore Sturgeon's novels *More Than Human* and *The Cosmic Rape*, and the Hammer film *Village of the Damned*, presumes the ability to remotely link neuronal activity for various individual agents. Certainly, the apparatus for a functional MRI device is considerably larger than the compact helmets for "mind reading" frequently depicted in science fiction magazines and comic books, and this technique provides information only about blood flow in the brain. While many remain unconvinced, some believe that this technique may someday scrve as an accurate lie detector, enabling us to directly discern a person's thoughts and intentions. Quantum mechanics brought us a world unimagined by the science fiction stories of fifty years ago, and it may now actually start bringing aspects of the science fictional world into reality.

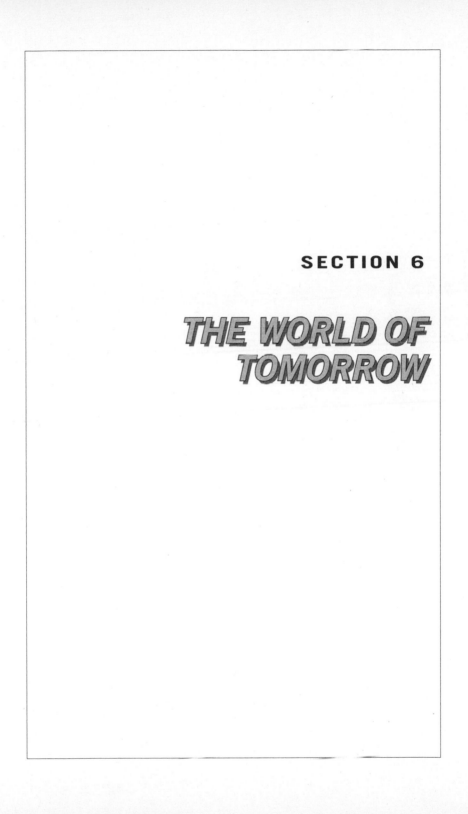

SECTION 6

THE WORLD OF TOMORROW

CHAPTER TWENTY

Coming Attractions

I have described the basic concepts underlying
quantum mechanics and have discussed how these principles account for the properties of single atoms, nuclei, and many-body systems, such as metals and semiconductors. By elucidating the physics
of the laser, the diode, the transistor, and disc drives, we now have
an understanding of the basic building blocks of such modern technology as laptops, DVDs, and cell phones, which, for many, are part
of everyday life in the twenty-first century. All of us routinely make
use of devices and applications that would not be possible without
the understanding of nature provided by quantum mechanics.

We have obviously not gone into any detailed descriptions as
to how consumer products, such as a computer, operate. By combining the electrical-current inputs in a series of transistors and
diodes in ingenious ways, one can arrange it so that two high currents cancel each other out and lead to either a low current (two
"ones" combine to form a "zero") or a high current (two "ones"
yield another "one"). Similarly, if one current level is high and the
other is low, then a circuit can be constructed so that the output is
either high or low, depending on the required logic operation. In
this way the "ones" and "zeros" in the computer can be manipulated. A full discussion of Boolean mathematics, logic gates, and
data storage and processing employed in a computer would be
fascinating (in my nerdy opinion), but it would involve no new
principles or applications of quantum mechanics.

Nevertheless, I would like to note in passing that sometimes the physics from Section 3 (radioactivity) interferes with the physics of Section 5 (solid state devices). Few have not experienced the frustration of having a computer program freeze or crash for no particular reason, solved only by a rebooting of the operating system. Sometimes the source of the problem turns out to be thorium, a radioactive element that is a contaminant (thorium is as common as lead) in the circuitry packaging. When the thorium nucleus decays it emits an alpha particle and the high-energy helium nucleus can disrupt the current in an unlucky transistor. The loss of information in the middle of a calculation often requires that the entire program be restarted from scratch. What quantum mechanics giveth, quantum mechanics taketh away.

Similarly, the scheme by which mobile phones send and receive electromagnetic signals to radio towers that then connect to a land-based call router is rather clever. Present models use very sophisticated protocols to determine the optimal region, or "cell," with which to transmit the call, but the basic quantum mechanics in the cell phone itself essentially involves applications of transistors and diodes. The heart of the cordless phone is the analog-to-digital converter, which takes a variable voltage generated when spoken sound waves are transformed into electrical signals (as occurs in a conventional landline phone) and, through what is effectively an operational amplifier (a series of transistors and resistors), converts this variable voltage to either a "one" or a "zero." Obviously, devices must also exist that operate in reverse, so that a digital-to-analog conversion can transform the received signal into the variable sound we hear.

One really cool feature in the latest generations of cell phones is the touch screen. While many simple touch screens, as in kiosk information displays or automatic teller machines, detect the alteration in electric field caused by the electrical conductance and capacitance of your finger, the newest multitouch versions send an LED-generated infrared light beam skimming along the inner surface of the screen. Touching the screen causes some of the infrared light to be scattered, and the location of your finger is ascertained by determining which photodetector behind the screen collects the scattered light. Here again, all the quantum physics in a touch screen is represented in the infrared light-emitting diodes and photodetectors.

Consequently, rather than delve into the inner workings of a variety of electronic products, which will not necessarily add much to our discussion of quantum mechanics, I use this final section to describe how quantum physics may continue to shape the future. That is, I would like to describe the concepts and devices that may become part of our world five or ten or twenty years from now. I will not try to make actual predictions, as that is a mug's game, but will rather explain the relevant quantum mechanics that underlies such phenomena as "quantum computers" and "nanotechnology." We already understand the basics; now I will discuss some novel advanced applications that may be coming to a consumer electronics store near you the day after tomorrow.

As described in Chapter 18, the greater sensitivity of read-head sensors using the giant magnetoresistance effect and magnetic tunnel junctions means that smaller magnetic bits can be detected, resulting in an increase in the storage capacity of hard drives. These magnetic sensors can also enable faster data retrieval. The polarization and depolarization of the magnetic layers in the sensor happens quickly, so the detector can read the bits even when the hard-drive platters are rotating at speeds of over ten thousand revolutions per minute. But the next generation of "semiconductor spintronic" devices may increase the speed of computers even more, by removing the need for a separate magnetic storage medium.

There has been considerable interest by researchers in developing semiconductor transistor structures that make use of the electron's internal magnetic field to process information. Using magnetic metals as the electrodes on a semiconductor device, it is possible to inject magnetically polarized electrons, that is, charge carriers whose internal magnetic fields all point in the same direction, into a semiconductor. By varying the magnetic field in the semiconductor device, the current could be controlled without the need to change the concentration of charge carriers, as in the field-effect transistor discussed in Chapter 17. The goal is to construct a device with a steeper on-off transition for the high-low current levels that are used to represent "ones" and "zeros," with faster switching between these two states that uses less energy to operate.

This last point is important. Each transistor in your computer creates a small amount of heat as it drives a current from low to high values and back again (passing a current through a toaster wire or lightbulb filament generates heat, and the same physics applies inside a semiconductor transistor). When millions and millions of these transistors are packed into a confined space, the resulting temperature rise can be significant, and in turn this consideration can limit the integrated circuit's performance. Hence the need for multiple cooling fans in most computer towers. "Spin transistors" use less power, thereby allowing a greater number of devices to be placed in close proximity, resulting in more computing power packed onto a microprocessor.

In addition, the ability to both store magnetic information and manipulate ones and zeros as in an integrated circuit suggests that it might be possible to combine both magnetic data storage and computer logic functions on a single chip. In the late 1950s, when the Challengers of the Unknown faced off against a "calculating machine" capable of independent thought, it was believed that such a device would have to be the size of a large room, while in the future, thanks to quantum mechanics, we may all be able to carry our own ULTIVACs in our back pockets.

The need for speed in computation is driving interest in an even more exotic use of quantum mechanical spin in calculators, often referred to as "quantum computers." Of course, in a sense *all* computers (aside from an abacus or a slide rule) are quantum computers, in that their fundamental data-processing elements, diodes and transistors, would not have been invented if not for the insights into the properties of matter provided by quantum theory.

A "quantum computer" is a different beast entirely. In short, rather than represent a "one" or a "zero" through a high or low current passing through a transistor, or from a region of magnetic material with its north pole pointing in one direction or the other, the atoms themselves are the ones and zeros. Actually, quantum computers have been proposed that involve atoms, nuclei, ions, photons, or electrons as the basic computing element—I focus my discussion on electrons for simplicity. Electrons have an intrinsic angular momentum of either $+\hbar/2$ or $-\hbar/2$, and in a quantum computer these are the elements that will represent the ones and zeros.

While using electrons in this way would indeed shrink the size of the computer's elements nearly as far as physically possible, this

alone is not what motivates research in quantum computers. Proposed quantum computers involve using pairs of identical particles,
arranged so that their wave functions overlap. Recall the discussion from Chapter 12 of two electrons brought so close that their de
Broglie waves interfered. In this case we represented the new two-
electron wave function by a single ribbon, where one side of the
ribbon was white and the other was black. In the example in Figure
30, both sides of the ribbon facing out were white. But we could also
have held both black sides facing out, or the left side could have
been black and the right side white, or the reverse. To represent
these four possibilities—white, white; black, black; black, white;
white, black—using conventional computer elements would require
two transistors, and they could generate these states only one at a
time, that is, in series.

With the quantum ribbon from Chapter 12, all four states are
possible simultaneously—if the ribbon is in a dark room and we
don't know which sides are facing out. In this case, all four states
may be present, and until we turn on the lights and examine the
ribbon, the ribbon can represent the four possible states in parallel.
Where I need four separate conventional bits to represent these outcomes, I need only two "quantum bits," or "qubits," to accomplish
the same task. While the ribbon analogy breaks down when dealing
with more than two entangled wave functions, the arguments hold,
and one needs only three quantum bits to represent eight distinct
conventional states, and ten qubits can do the work of 1,024 classical bits.

This parallelism implies that a quantum computer could perform calculations must faster than a conventional computer. En-
cryption of information for national security, online commerce, or
just using a credit card to pay for a purchase at a gas station involves knowledge of the prime-number factors* of numbers that
are so large that even the fastest conventional computers could not
determine the factors in a reasonable time. Quantum computers,

* A prime number is a number that can be divided only by itself or 1. The number 5
is a prime number, while 8 is not (it can be divided by 2 and 4, in addition to 8 and 1).
Computer security involves numbers that are the product of two very large prime
numbers. Security is maintained by requiring both sides of the transaction to know
the factors (which are so large that it is impossible to guess even using conventional
supercomputers to try all possibilities).

with the ability to perform multiple tasks at the same time, could change this situation. A small-scale prototype quantum computer has been able to successfully factor a two-digit number ($15 = 5 \times 3$), but a fully operational quantum computer does not exist and is years and years away in the most optimistic scenario. Nevertheless, data security and cryptology will be dramatically changed if such devices are ever constructed. It will be up to all of us to ensure that this technology does not fall into the wrong hands, for who could forget when the evil Decepticons used quantum computers to hack into the Pentagon's secure computer system in the *Transformers* movie (2007), cracking a code in ten seconds that would take more than two decades for the most power supercomputer.

Now, to say that two overlapping electrons can represent all four spin combinations, provided we don't examine them, may seem a bit* of a cheat—you can argue that any pair of transistors can represent all four states if I do not actually examine whether the currents passing through them are high or low. But there is a fundamental difference in the quantum spin case that gets to the heart of some of the philosophical arguments over the role of measurement in quantum mechanics.

Throughout this entire book I have pulled a fast one on you, Fearless Reader, and its time to come clean. The difference between the quantum case of overlapping electronic wave functions and the conventional situation involving transistors, and the reason that the quantum ribbon can represent all four possible outcomes simultaneously, is that if the ribbon remains in the dark, that is, until I do a measurement and examine it, the very concept of the color of the ribbon is not well defined.

Let's return to the real world of electrons for a moment. I have specified that an electron can have only one of two possible values of its intrinsic angular momentum, rotating either clockwise (that is, spin "up") or counterclockwise (spin "down"). But we never asked the question: Up or down—relative to what? Clockwise or counterclockwise rotations—about what axis?

The intrinsic angular momentum has a small magnetic field associated with it, with a north pole and south pole. If I don't measure this magnetic field, that is, in the absence of an external mag-

---

* No pun intended.

netic field, then I have no way of knowing where the pole is oriented. If I apply an external magnetic field and measure the electron's magnetic orientation, it will either line up with the field or be 180 degrees opposite to the external field (as in our discussion of MRI in Chapter 19). If the external magnet I apply has its north pole pointing toward the ceiling of the room you are in, then this defines the "up"/"down" direction. The electron's magnetic field will point either to the ceiling or to the floor. If the external magnetic field is instead applied pointing toward one of the walls in your room, then *this* defines the "up"/"down" direction, and the electron's magnetic field will point either toward the wall or toward the wall opposite it. Once I apply an external field and measure the electron's magnetic field, that very act defines the axis about which "clockwise" and "counterclockwise" make sense, and until I do, all I can say is that the electron is in some superposition of these possibilities. It is in this way that we can say that a quantum system can represent multiple states simultaneously.*

Einstein smelled a rat in this scenario and spent a considerable fraction of his later years trying to catch it, for the situation just described opens up the possibility for information to travel faster than the speed of light. Say I arrange two electrons so that their wave functions overlap, and they can be described by a two-particle wave function (as in Chapter 12) and with total intrinsic angular momentum together to be zero. One electron has spin = $+\hbar/2$ and the other has spin = $-\hbar/2$, but let's say I don't know which is which. When I measure the electron on the left and find that its spin is $+\hbar/2$, not only do I know logically that the other spin must be $-\hbar/2$ (since I already knew that the total spin was zero), but I also now know *what axis* the second electron will be anti-aligned with! The process of measuring the first electron's magnetic field picks a preferred direction not only for that electron, but for the other one as well, since they are both part of the same wave function, which contains all the information about the system.

Now, here's where it gets fun. Let's assume I have an infinitely stretchy ribbon representing the two electrons. I hold one end of the ribbon and pull the other end all the way across town, keeping the

---

* This is similar to the question of position and momentum measurements in Chapter 7. The answer you receive depends on what question you ask.

electrons still connected. Now I measure the spin of one electron, by placing it in an external magnetic field. This not only tells me if this electron is pointing with or against the applied field, but determines what direction the electron's magnetic field points. As soon as I do this, the properties of the other electron are also determined. After all, both electrons are described by a single wave function and thus behave as a single entity. In this way the entangled quantum state is like the famous twins in fairy tales, joined by a special bond, so that what happens to one is instantly felt by the other. Einstein objected that this would enable information (about the direction of the magnetic field being used in my lab to measure the electron) to be transmitted from one point in space to another, potentially faster than light could cover the same distance. In his famous phrase, this represented "spooky action at a distance," and he would have none of it.

Books have been written over the question of whether this scenario does indeed provide a mechanism for instantaneous transmission of information, and, if it does, how to reconcile this with the principles of the Special Theory of Relativity that states that nothing, not even information, can travel faster than light. It remains a topic of lively debate among physicists. As the man said when asked by the child about the nature of the afterlife—experts disagree. Now, questions of what "happens" to a quantum system in the moment of observation are fascinating, but a full discussion of these topics is not the focus of this text. I will therefore now exit, stage left, following two brief observations.

First, the issue of faster-than-light transmission of information holds *only* if the two electrons are described by a single wave function, and that will be true *only* if the "infinitely stretchy" ribbon does not break as I pull the two ends farther and farther apart. As you might expect, the more the ribbon is stretched, the easier it is for some stray perturbation to disturb the overlapping waves between the two ends. Once the connection between the two ends is severed, then a measurement of the spin of one electron will have no bearing at all on the other electron, as they are now described by two distinct ribbons. The fancy way to describe this is that the two electrons' wave functions must remain "entangled" in order for this process to hold, and any object or input of energy that dis-

Figure 50: *The Atom (who in his secret identity is physics professor Ray Palmer), informs his Justice League of America colleagues about recent experiments involving entangled quantum states of photons in JLA # 19. Professor Palmer is referring to the work of Anton Zeilinger and coworkers, published in* Physical Review Letters *(1998). The second panel is an illustration of the overinterpretation of the principles of quantum mechanics to matters of spirituality—we will avoid this trap, as we do all such traps into which the Justice League may fall.*

turbs this state (breaks the ribbon) causes "decoherence." Overcoming the enormous challenges involved in avoiding decoherence keeps experimental physicists busy, and whether a functioning quantum computer is ever constructed that can live up to its potential remains to be seen.

The second point is that whenever one reads in the popular science press about recent experiments in "teleportation," what they are always referring to is the transmission of information concerning a quantum state, similar to the situation described above. They are not dealing with "beaming" people as in *Star Trek*, or sending atoms or electrons from one point in space to another. You will have to face your daily morning commute for quite some time.

There have recently been experiments that indeed support the notion that information concerning two entangled quantum entities, such as the polarization of photons, can be transmitted even when separated by a great distance. That is, experimental techniques have now advanced such that considerations that previously had been purely theoretical may be put to empirical verification, as shown in Figure 50, from a 1998 issue of the adventures of the Justice League of America. Real science is now inspiring the comic books, and not the other way round!

CHAPTER TWENTY-ONE

Seriously, Where's My Jet Pack?

As mentioned at the very start of our narrative, science fiction pulp writers expected that the future would bring a new era in energy production and storage. Instead, it was data manipulation that underwent a profound transformation, enabled by the discoveries of quantum mechanics.

Why is a new type of energy-delivery system needed before jet packs and flying cars become commercially viable? Let's stipulate that we are not invoking any violations of the laws of physics, such as the discovery of "cavorite" or some other miraculous material with antigravity properties. Thus, our jet pack must provide a downward thrust, equal to a person's weight, in order to lift the person off the ground. Consider how much energy it takes to lift a 180 pound person 330 feet (one-sixteenth of a mile) up in the air. Just to get up there, neglecting any energy needed to jet from place to place, would require an energy expenditure of a little over eighty thousand Joules, which is equivalent to 0.5 trillion trillion electron Volts.

Recall that nearly every chemical reaction involves energy transfers on the order of an electron Volt. Thus, to lift a person over a twenty-story building involves roughly a trillion trillion molecules of fuel. But that's not actually as much as it seems, for there are approximately that many atoms in twenty cubic centimeters of any solid (recall that a cubic centimeter is about the size of a sugar cube). A gallon of fuel contains nearly four thousand cubic centimeters, capable of producing over a hundred trillion trillion electron Volts of energy. If this is the case, why are we still driving to work?

The problem is—what goes up must come down. As soon as our jet pack stops expending energy to maintain our large potential energy above ground, back to Earth we return. Thus, every second we spend in the air, we must continue to burn through our stored chemical energy. The rate at which we use up fuel will depend on the particular mechanism by which we achieve an upward thrust, but for most energy supplies, our trip will be over in a minute or two. We can indeed take jet packs to work, provided we live no more than a few blocks from our office.

Note that the largest expenditure in energy is overcoming gravity. It takes more than eighty thousand Joules to get us up in the air. Flying at forty miles per hour, in contrast, calls for a kinetic energy of only thirteen thousand Joules (neglecting the work we must do to overcome air resistance). *This* is why we don't have flying cars. Your gas mileage would be nonexistent if the vast majority of the fuel you carried went toward lifting you up off the ground, with hardly any left over to get you to your destination (sort of defeats the whole purpose of a car, flying or otherwise).

There have been improvements in the energy content of stored fuel, and prototype jet packs have been able to keep test pilots aloft for more than a minute, but ultimately, the longer the flight, the more fuel needed (and the heavier the jet pack will be). Of course, there are alternatives to chemical-fuel reactions to achieve thrust and lift. One could use a nuclear reaction, which, as we saw in Section 3, yields roughly a million times more energy per atom than chemical combustion, but the idea of wearing even a licensed nuclear power plant on your back is less than appealing (except possibly for the Ghostbusters).

In the 2008 film *Iron Man*, Tony Stark designs a suit of armor that contains a host of high-tech gadgets, all of which are within the realm of physical plausibility—with one big exception. The one miracle exception from the laws of nature that the film invokes is the "arc reactor" that powers Stark's high-tech exoskeleton. This device is a cylinder about the size of a hockey puck and is capable of producing "three GigaWatts of power,"* sufficient to keep a real-world jet pack aloft and flying for hours. Sadly, we have no way of producing such compact, lightweight, high-energy-content power cells.

Had the revolution in energy anticipated by the science fiction pulp magazines indeed occurred and we employed personal jet packs to get to work or the corner grocery store, powered by some exotic energy source, the need for conventional fossil fuels would of course be dramatically reduced, with a concurrent dramatic shift in geopolitical relations. There is one important use of potential jet-pack technology that does not involve transportation but rather thirst quenching, that would have an immediate beneficial impact.

According to the World Health Organization, as of 2009, 40 percent of the world's population suffers from a scarcity of potable fresh water. The most straightforward method to convert seawater to fresh water involves boiling the salt water and converting the liquid water to steam, which leaves the salts behind in the residue. This is, after all, what occurs during evaporation from the oceans, which is why rainwater is salt free. The amount of energy needed to boil a considerable amount of water is not easily provided by solar cells, but if one had a power supply for a fully functioning jet pack, the lives of more than two billion people would be profoundly improved, even if everyone's feet stayed firmly planted on the ground.

Can quantum mechanics help in the production of energy, so that the jet-pack dreams of the 1930s can be at long last realized? Possibly. Global consumption of energy, which in 2005 was estimated to be sixteen trillion Watts, will certainly increase in the future, with many experts projecting that demand will grow by nearly 50 percent in the next twenty years. One strategy to meet this additional need involves the construction of a new power plant, capable of producing a gigaWatt of power, at the rate of one new facility every day for the next two decades. This does not seem likely to happen.

---

* This is for the first version of the arc reactor, "built in a cave—with a bunch of scraps!" Later models had even higher power outputs, though the exact specs are the classified proprietary information of Stark Enterprises.

Another approach is to tap the vast amount of energy that is, for the most part, ignored by all nations—sunlight. The surface of the Earth receives well over a hundred thousand trillion Watts of power, more than six thousand times the total global energy usage and more than enough to meet the world's energy needs for decades to come. As described in Chapter 16, the simple diode, comprised of a junction between one semiconductor with impurities that donate excess electrons and a second semiconductor with impurities that donates holes, can function as a solar cell. When the diode absorbs a photon, an electron is promoted into the upper band, leaving a mobile hole in the lower filled band. These charge carriers feel a force from the strong internal electric field at the pn junction, and a current can be drawn out of the device, simply as a result of exposing it to sunlight. Work is under way to improve the conversion efficiency of these devices—that is, to maximize the current that results for a given intensity of sunlight. But even using current cells, with conversion efficiencies of only 10 percent (that is, 90 percent of the energy that shines on the solar cell does not lead to electrical power), we could provide all the electricity needs of the United States with an array of solar cells of only one hundred miles by one hundred miles.

The problem is, we don't have enough solar cells on hand to cover a one-hundred-mile-by-one-hundred-mile grid, and at the present production capacity it would take many years to fabricate these devices. Moreover, even if the solar cells existed, we would need to get the electrical power from bright sunny locales to the gloomy cities with large population densities. Here again, quantum mechanics may help.

In Chapter 13 we saw that at low temperatures certain metals become superconductors, when their electrons form bound pairs through a polarization of the positive ions in the metal lattice. Electrons have intrinsic angular momentum of $\hbar/2$ and individually obey Fermi-Dirac statistics (Chapter 12) that stipulate that no two electrons can be the same quantum state. When the electrons in a metal at low temperature pair up, they create composite charge carriers that have a net total spin of zero. These paired electrons obey Bose-Einstein statistics, and as the temperature is lowered they condense into a low energy state. If the temperature of the solid is low enough, then for moderate currents there is not enough energy to scatter the electrons out of this lowest energy state, and they can thus carry current without resistance. This phenomenon—superconductivity—is an intrinsically quantum mechanical effect and is observed only in metals at extremely low temperatures, below –420 degrees Fahrenheit.

At least—that was the story until 1986. In that year two scientists, Johannes Bednorz and Karl Müller, at the IBM research laboratory in Zurich, Switzerland, reported their discovery of a ceramic that became a superconductor at –400 degrees Fahrenheit. That's still very cold, but at the time it set a record for the highest temperature at which superconductivity was observed. Once the scientific community knew that this class of materials, containing copper, oxygen, and rare Earth metals, could exhibit superconductivity, the race was on, and research labs around the world tried a wide range of elements in a host of combinations. A year later a group of scientists from the University of Houston and University of Alabama discovered a compound of yttrium, barium, copper, and oxygen that became a full-fledged superconductor at a balmy –300 degrees Fahrenheit. Liquid nitrogen, used in many dermatologists' offices for the treatment of warts, is 20 degrees colder at –321 Fahrenheit. These materials are referred to as "high-temperature superconductors," as their transition into a zero resistance state can be induced using a refrigerant found in many walk-in medical clinics. There is no definitive explanation for how these materials are able to become superconductors at such relatively toasty temperatures, and their study remains an active and exciting branch of solid-state physics. The most promising models to account for this effect invoke novel mechanisms that quantum mechanically induce the electrons in these solids to form a collective ground state.

High-temperature superconductors would be ideal to transmit electricity generated from a remote bank of solar cells or windmills to densely populated regions where the power is needed. While it would need to be kept cool, liquid nitrogen is easy to produce, and when purchased for laboratory needs it is cheaper than milk (and certainly cheaper than bottled water). Unfortunately, to date challenging materials-science issues limit the currents that can be carried by these ceramics, such that if we were to use them for transmission lines they would cease to become superconductors and would in fact have resistances higher than those of ordinary metals.

If these problems are ever solved, then along with transmitting electrical power, these innovations may help transportation undergo a revolution as well. As discussed in Chapter 13, in addition to carrying electrical current with no resistance, superconductors are perfect diamagnets, completely repelling any externally applied magnetic field. The material sets up screening currents that cancel out the external field trying to penetrate the superconductor, and as there is no resistance to current flow, these screening currents can persist indefinitely. If high-temperature superconductors can be fabricated that are able to support high enough currents to block out large enough magnetic fields, then high-speed magnetically levitating trains are possible, where the major cost involves the relatively cheap and safe liquid nitrogen coolant.

Bednorz and Müller won the Nobel Prize in Physics just one year after they published their discovery of high-temperature superconductivity in ceramics. However, more than twenty years later, the trains still do not levitate riding on rails composed of novel copper oxide compounds. Unlike giant magnetoresistance and the solid-state transistor, both of which went from the research lab to practical applications in well under a decade, there are no preexisting consumer products for which raising the transition temperature of a superconductor would make a significant difference. Nevertheless, research on these materials continues, and someday we may have high-temperature superconductors overhead in our transmission lines and underfoot on our rail lines.

Another untapped source of energy that quantum mechanics–based devices may be able to exploit in the near future involves waste. I speak here not of garbage but of waste heat, generated as a by-product of any combustion process.

Why is heat wasted under the hood of your car? Heat and work are both forms of energy. Work, in physics terms, involves a force applied over a given distance, as when the forces exerted by the collisions of rapidly moving gas molecules lift a piston in a car engine. Heat in physics refers to the transfer of energy between systems having different average energy per atom. Bring a solid where the atoms are vigorously vibrating in contact with another where the atoms are slowly shaking, and collisions and interactions between the constituent atoms result in the more energetic atoms slowing down while the sluggish atoms speed up. We say that the first solid initially had a higher temperature while the second had a lower temperature, and that through collisions they exchange heat until they eventually come to some common temperature. We can do work on a system and convert all of it to heat, but the Second Law of Thermodynamics informs us that we can never, with 100 percent efficiency, transform a given amount of heat into work.

Why not? Because of the random nature of collisions. Consider the molecules in an automobile piston, right before the ignition spark and compression stroke cause the gasoline and oxygen molecules to undergo combustion. They are zipping in all directions, colliding with each other and the walls and bottom and top of the cylinder. The pressure is uniform on all surfaces in the cylinder. Following combustion, the gas-oxygen mixture undergoes an explosive chemical reaction, yielding other chemicals and releasing heat; that is, the reaction products have greater kinetic energy than the reactants had before the explosion. This greater kinetic energy leads to a greater force being exerted on the head of the piston as the gas molecules collide with it. This larger force raises the piston and, through a clever system of shafts and cams, converts this lifting to a rotational force applied to the tires. But the higher gas pressure following the chemical explosion pushes on all surfaces of the cylinder, though only the force on the piston head results in useful work. The other collisions wind up warming the walls and piston of the cylinder, and from the point of view of getting transportation from the gasoline, this heat is "wasted."

When heat is converted to work, the Second Law of Thermodynamics quantifies how much heat will be left over. In an automobile, in the best-case scenario, one can expect to convert only one-third of the available chemical energy into energy that moves the car, and very few auto engines are even that efficient. There's a lot of energy under the hood that is not being effectively utilized. Similarly, cooling towers for power plants eject vast quantities of heat into the atmosphere. It is estimated that more than a trillion

Watts of energy are wasted every year in the form of heat not completely converted to work. This situation may change in the future, thanks to solid-state devices called "thermoelectrics." These structures convert temperature differences into voltages and are the waste-heat version of solar cells (also known as "photovoltaic" devices) that convert light into voltages.

Thermoelectrics make use of the same physics that enables solid-state thermometers to record a temperature without glass containers of mercury. Consider two different metals brought into contact. We have argued that metals can be viewed as lecture halls where only half of the possible seats are occupied, so that there are many available empty seats that can be occupied if the electrons absorb energy from either light, or applied voltages, or heat. Different metals will have different numbers of electrons in the partially filled lower band. Think about two partially filled auditoriums, each with different numbers of people sitting in the seats, separated by a removable wall, as in some hotel ballrooms. One auditorium has two hundred people, while the other has only one hundred. Now the wall separating them is removed, creating one large auditorium. As everyone wants to sit closer to the front, fifty people from the first room move into vacant seats in the other, until each side has one hundred and fifty people sitting in it. But both metals were electrically neutral before the wall was removed. Adding fifty electrons to the small room creates a net negative charge, while subtracting fifty electrons from the first room yields a net positive charge. A voltage thus develops at the juncture between the two metals, just by bringing them into electrical contact. If there are significant differences in the arrangements on the rows of seats in each side, then as the temperature is raised the number of electrons on each side may vary, leading to a changing voltage with temperature. In this way, by knowing what voltage measured across the junction corresponds to what temperature, this simple device, called a "thermocouple," can measure the ambient temperature.

Thermoelectrics perform a similar feat using a nominally homogenous material, typically a semiconductor. If one end of the solid is hotter than the other, then the warmer side will have more electrons promoted from the full lower band up into the mostly empty conducting band than will be found at the cooler end. For some materials the holes that are generated in the nearly filled lower-energy orchestra will move much slower than the electrons in the higher-energy balcony, so we can focus only on the electrons. The electrons promoted at the hot side will diffuse over to the cooler end, where they will pile up, creating a voltage that repels any additional electrons from moving across the semiconductor. This voltage can then be used to run any device, acting as a battery does. To make an effective thermoelectric device, one wants a material that is a good conductor of electricity (so that the electrons can easily move across the solid) but a poor conductor of heat (so that the temperature difference can be maintained across the length of the solid). Research in developing materials well suited to thermoelectric applications is under way at many laboratories. Commercially viable devices could find application in, for example, hybrid automobiles, taking the waste heat from the engine and converting it into a voltage to charge the battery. In the world of the future, thanks to solid-state devices made possible through our understanding of quantum mechanics, the cars may not fly, but they may get much better mileage.

Another way to extract electrical power from random vibrations involves nanogenerators. These consist of special wires only several nanometers in diameter, composed of zinc oxide or other materials that are termed "piezoelectric." For these compounds a mechanical stress causes a slight shift in the crystal structure, which then generates a small voltage. Progress has been made in fabricating arrays of nanoscale wires of these piezoelectric materials. Any motion or vibration will cause the tiny filaments to flex and bend, thereby creating an electric voltage that can be used to provide power for another nanoscale machine or device.

Finally, we ask, can quantum mechanics do anything to develop small, lightweight batteries to power a personal jet pack? The answer may lie in the developing field of "nanotechnology." "Nano" comes from the Greek word for "dwarf," and a nanometer is one billionth of a meter—equivalent to approximately to the length of three atoms placed end to end. First let's see how normal batteries operate, and then I'll discuss why nanoengineering may lead to more powerful energy-storage devices.

In an automobile engine the electrical energy from the spark plug induces the chemical combustion of gasoline and oxygen. Batteries employ a reverse process, where chemical reactions are used to generate voltages.

In an electrolysis reaction, an electrical current passes through reactants (often in liquid form) and provides the energy to initiate a chemical reaction. For example, one way to generate hydrogen gas (that does not involve the burning of fossil fuels) is to break apart water molecules. To do this we insert two electrodes in a beaker of water and attach them to an external electrical power supply, passing a current through the fluid. One electrode will try to pull electrons out of the water (pure water is a very good electrical insulator), while the other will try to shove them in. The input of electrical energy overcomes the binding energy holding the water molecule together, and positively charged hydrogen ions (H+) are attracted to the electrode trying to give up electrons, while the negatively charged hydroxides (OH– units) move toward the electrode trying to accept electrons. The net result is that H_2O molecules break into gaseous hydrogen and oxygen molecules.

In a battery, making use of essentially a reverse electrolysis process, different metals are employed for the electrodes (such as nickel and cadmium); they are chosen specifically because they undergo chemical reactions with certain liquids, leaving the reactant either positively or negatively charged. Where the metal electrode touches the chemical fluid (though batteries can also use a porous solid or a gel between the electrodes), electrical charges are either taken from the metal or added to it, depending on the chemical reaction that proceeds.* A barrier is placed between the two electrodes, preventing the fluid from moving from one electrode to the other, so that negative charges (that is, electrons) pile up on one electrode and an absence of electrons (equivalent to an excess of positive charges) accumulates at the other.

The only way the excess electrons on one electrode, which are repelled from each other and would like to leave the elec-

---

* Just as, in our discussion of solid-state thermometers a moment ago, the metal electrode and the contacting fluid represent two partially filled auditoriums, which will have electrons move from one room to the other, depending on which room has the higher concentration of electrons.

trode, can move to the positively charged electrode is if a wire is connected across the two terminals of the battery. The stored electrical charges can then flow through a circuit and provide the energy to operate a device. In an alkaline battery, once the chemical reactants in the fluid are exhausted, the device loses its ability to charge up the electrodes. Certain metal-fluid chemical reactions can proceed in one way when current is drawn from the battery, and in the reverse direction with the input of an electrical current (as in the water electrolysis example earlier), restoring the battery to its original state. Such batteries are said to be "rechargeable," and it is these structures that have exhibited the greatest increases in energy-storage capacity of late.

There have been great improvements in the energy content and storage capacity of rechargeable batteries, driven by the need for external power supplies in consumer electronics. In a battery the electrodes should be able to readily give up or accept electrons. Examination of the periodic table of the elements shows that lithium, similar in electronic structure to sodium and hydrogen, has one electron in an unpaired energy level (shown in Figure 31c) that it easily surrenders, leaving it positively charged. Batteries that make use of these lithium ions, with a lithium-cobalt-oxide electrode,* and with the other electrode typically composed of carbon, produce nearly twice the open-circuit voltage of alkaline batteries. These batteries are lighter than those that use heavy metals as the electrodes, and a lithium-ion battery weighing eight ounces can generate more than 100,000 Joules of energy, compared to 50,000 Joules from a comparable-weight nickel-metal-hydride battery or 33,000 Joules from a half-pound lead-acid battery. These lightweight, high-energy-capacity, rechargeable batteries are consequently ideal for cell phones, iPods, and laptop computers.

As all the electrochemical action in a battery takes place when the electrolyte chemical comes onto physical contact with the electrode surface, the greater the surface area of the electrode, the more available sites for chemical reactions to proceed. One way to increase the surface area is to make the electrodes larger, but this

---

* Though other chemical compounds are employed depending on the battery requirements.

conflicts with the desire for smaller and lighter electronic devices. Another way to increase the capacity of these batteries is to structure the electrodes differently. Nanotextured electrodes are essentially wrinkly on the atomic scale, dramatically increasing the surface area available for electrochemical reactions without a corresponding rise in electrode mass. Recent research on electrodes composed of silicon nanoscale wires finds that they are able to store ten times more lithium ions without appreciable swelling than carbon electrodes can. While not quite in the league of Iron Man's arc reactor, the ability to fabricate and manipulate materials on these nanometer-length scales is yielding batteries with properties worthy of the science fiction pulps.

This nanostructuring is also helping out with the laundry. Nanoscale filaments woven into textiles yield fabrics that are wrinkle resistant and repel staining. In addition to giving us whiter whites, nanotechnology is helping keep us healthy. A five-nanometer crystal contains only thirty-three hundred atoms, and such nanoparticles are excellent platforms for highly refined pharmaceutical delivery systems, able to provide, for example, chemotherapy drugs directly to cancerous cells while bypassing healthy cells.

We are only beginning to exploit the quantum mechanical advantages of nanostructured materials. There are ninety-two stable elements in the periodic table, and the specific details of the configuration of their electrons determines their physical, optical, and chemical properties. Crystalline silicon has a separation between its lowest filled states and first empty states of about one electron Volt, and if you want a semiconductor with a different energy gap, you must choose a different chemical element. Many technological applications would become possible, or would be improved, if the energy separation in crystalline silicon could be adjusted at will, without alloying with other chemicals that may have unintended deleterious effects on the material's properties. Recent research indicates that we can indeed make silicon a "tunable" semiconductor, provided we make it tiny.

Whether the energy separation between the filled orchestra and the empty balcony in our auditorium analogy is one electron Volt (in the infrared portion of the spectrum), two electron Volts (red light), or ten electron Volts (ultraviolet light) is determined by the elements that make up the solid and the specific details of how each atom's quantum mechanical wave function overlaps and interacts with its neighbors. In large crystals, big enough to see with the naked eye or with an optical microscope, the electrons leaving one side of the solid will suffer many scattering collisions, so any influence from the walls of the crystal on the electron's wave function will have been washed out by the time the electron makes it to the other side. If the size of the solid is smaller than the extent of the electron's de Broglie wavelengths, then the electrons in the small crystal in essence are able to detect the size of the solid in which they reside. The smaller the "box" confining these electrons, the smaller the uncertainty in their location and—thanks to Heisenberg—the larger the uncertainty in their momentum. Consequently, "nanocrystals" can have an energy band gap that is determined primarily by the size of the solid and that we can control, freeing designers of solid-state devices from the "tyranny of chemistry."

The discoveries by a handful of physicists back in the 1920s and 1930s, explicating the rules that govern how atoms interact with light and each other, continue to shape and change the world we live in today and tomorrow.

Journey into Mystery

Every morning when I look out the window, I am reminded that we do not live in the world promised by science fiction pulp magazines, as I note in the skyline the absence of zeppelins. However, before I arise, the programmable solid-state timer on my coffee maker begins brewing my morning java. I thus literally do not get up in the morning without enjoying the benefits of a world informed by quantum mechanics.

As noted in the introduction, the pulps and science fiction comic books of fifty years ago certainly missed the mark (sometimes by a wide margin) in their prognostications of the technological innovations we would enjoy in the far-off future of the twenty-first century, a chronological milestone we have now reached. While their crystal balls may have been foggy, these errors concerning technology seem presciently accurate compared to how far off their *sociological* predictions were. For example, few science fiction writers in the 1950s anticipated how much public and private space would be designated smoke-free, and it was generally expected that in the year 2000, as in the year 1950, a universal truth would remain that all scientists smoke pipes.

Predicting the evolution of language is another challenge for those trying to create visions of life in the distant future. It is amusing to read old Buck Rogers newspaper strips from 1929 on, and see, amid the descriptions of rockets ships, disintegration rays, and levitation belts, that colloquial expressions of early-twentieth-century

America remain vibrant and comprehensible in the twenty-fifth century. Buck, who fell asleep in the 1920s and awoke five hundred years later, can be excused for his use of slang, but apparently everyone in the future speaks this way. When facing an overwhelming robot army, warriors of a besieged city lament, "They've got us licked!" while another counsels, "Let's fade!" Gender equality appears set to move in reverse in the next five hundred years as well. Buck's fiancée leads a scouting team into enemy territory on Mars and gets separated from the rest of the group (cell phones appear to be a lost technology in the future). At the base ship Buck complains, "This is what comes of trusting a Woman with a Man's responsibility!" to which his lieutenant agrees, "They're all alike! They can drive a man crazy!" Just another reason why life as envisioned in science fiction isn't all it's cracked up to be.

Some of the writers of science fiction of fifty or more years ago had great optimism regarding the coming future. Those who were not proposing dystopian futures of atomic warfare and unceasing hostilities between nations (and alien species) were confident that many if not all of the ills that plague humankind would be defeated in the coming years thanks to . . . Science!

Science would free the housewife of the 1950s from the drudgery of housework and food preparation. The May 1949 issue of *Science Illustrated* speculated that a "New Wiring Idea May Make the All-Electric House Come True." The wiring idea involved dropping the operating voltage from 110 volts to 24 volts, using a small transformer.* The article argues that the benefit of using the lower voltage is that it enables the safe operation of many consumer items, and by adopting an "all-electric" household, the five-dollar cost of the transformer becomes a reasonable expense when amortized over a dozen household helpers. A photo spread shows that "a young housewife [. . .] from a single bedside panel with remote control switches [. . .] can turn on the percolator in the kitchen, turn radios on and off, light up a flood lamp in the yard for a late-home-coming husband [. . .] control the electric dishwasher and toaster [and] control every light and electrical outlet in and around a house

---

* These may seem like simple devices, but there is more to them than meets the eye!

from one single point." Little did the writer imagine that wireless technology, and semiconductor-based sensors whose operations could be preprogrammed, would remove the need for remote control switches on a bedside panel. No wonder social theorists worried about how the young housewife would fill the hours of the day in such a homemaker utopia.

Similarly, science has indeed revolutionized the workplace. Forget about inquiring, here in the twenty-first century, as to the location of our jet packs; what many want to know is: Where's our four-hour workday? It was a general expectation that by the year 2000, people would have so much leisure time that the pressing challenge would be to find ways to keep the populace entertained and occupied. Instead, for too many of us, the de facto workday has lengthened, thanks to the modern electronics that flowered from the development of quantum mechanics; the ability to be in constant contact has evolved into the necessity to be always connected.

Youngsters fifty years ago may not have been reading *Modern Mechanics* or *Popular Science,* but they learned of the brighter future to be delivered by scientific research and innovation in the pages of their comic books. While nowadays a best-selling comic book may have sales of a few hundred thousand copies, in 1960 sales of *Superman* comics were over eight hundred thousand per issue, and studies found that a single issue was shared and read by up to ten other kids. Lifelong attitudes about better living through technology were fostered in the four-color pages of these ten-cent wonders.

The Man of Tomorrow, in particular, starred in many classic stories describing the world of tomorrow. Superman was so popular in the 1940s and 1950s that at times he appeared in up to seven comics published by National Allied Periodicals (the company that would become DC Comics). In addition to his own stories in *Action* and *Superman,* the Man of Steel could be found in *Lois Lane, Superman's Girlfriend; Jimmy Olsen, Superman's Pal; World's Finest* (where he would team up every month with Batman and Robin); and *Superboy* and *Adventure Comics* (these latter two were filled with tales of Superman's teenage years as Superboy).

In 1958's *Adventure Comics* # 247, the Teen of Steel encounters three superpowered teenagers who, after playing some fairly harmless pranks on him, reveal to Superboy that they are from one thousand years in the future. These superteens have traveled back in time to offer Superboy membership in their club—the Legion of Superheroes. Apparently, one thousand years from now, a group of teenagers with a wide variety of powers and ability, from Earth and other planets, would band together to fight crime and evil throughout the United Planets. These young heroes were inspired by history tapes of the adventures of Superboy, and between their mastery of time travel and the Teen of Tomorrow's ability to fly so fast that he could "break the time barrier," Superboy would become a regular member of the Legion. Stories featuring the Legion would prove so popular with readers that they became a regular feature in *Adventure* and eventually squeezed Superboy out of the comic, aside from his appearances with the Legion in the thirtieth century.

According to these Legion tales, the promise of the space program and the race to the moon under way in the 1960s would culminate, in the thirtieth century, in a society ruled by and dedicated to science! In the world of the Legion of Superheroes, if you found yourself in trouble, you didn't call the police; you sent for the *science* police!

While evil despots and warlike alien races would still bedevil humanity in the year 2958, the Legion of Superheroes tales featured a general sense of progress and hope that may have, in part, accounted for their popularity. A thousand years hence, intelligence and knowledge would be honored and rewarded. In *Adventure* # 321, when Lightning Lad, one of the Legion's founding members, was sentenced to life in prison for "betraying" the Legion by "revealing" the secret of the Concentrator,* his cell featured buttons that when pressed would provide the *three* basic necessities of life: food, water, and . . . books! The writers of the Legion of Superheroes stories promised that in the future, we would live in a golden age of science.

Similarly, over at Marvel Comics (though in the mid- to late 1950s the company was known as Atlas), scientists were also given

---

* Don't worry, Fearless Reader—he was framed and eventually demonstrated his innocence.

pride of place in society. Stan Lee and Jack Kirby would not begin recounting the adventures of a quartet who took an ill-fated rocket trip "to the stars" and returned as the superpowered Fantastic Four until November 1961. Prior to this reintroduction of superheroes to the Marvel Comics universe in the 1960s, there were still plenty of menaces to be dealt with, as *Tales to Astonish, Amazing Fantasy, Strange Tales, Journey into Mystery,* and *Tales of Suspense* documented the near continuous onslaught of monstrous invaders from space, time, and other dimensions, all seeking global conquest. These would-be conquerors would regularly prove too much for local law enforcement and the military and often could be thwarted only by the lone efforts of a scientist!

And it's a good thing scientists were on the case, as Earth had to contend with the likes of Pildorr, Rorgg, Sporr, Orggo, Gruto, Rommbu, Bombu, Moomba, Dragoom, and Kraggoom. These creatures were rarely less than twenty feet tall and when not generic

Figure 51: Cover from Tales to Astonish # 13, describing the adventures of scientist Leslie Evans, who relates how "I Challenged Groot! The Monster from Planet X!" Such monstrous invaders from outer space threatened our planet several times a month in pre-superhero Marvel Comics.

bug-eyed monsters or monstrously oversized bugs, they were composed of stone, smoke, fire, water, electricity, wood, mud, or "ooz-
ing paint." But none of these were as fearsome as Orgg, the Tax
Collector from Outer Space!

Fairly typical was the November 1960 issue of *Tales to Astonish*
13 (Figure 51), where we hear the firsthand testimonial "I Chal-
lenged Groot! The Monster from Planet X!" This is presumably the
same Planet X that is home to Goom (and his son Googam); the
Thing from Planet X; and Kurrgo, the Master of Planet X.* Groot
was a giant treelike creature who came to Earth intending to steal
an entire village and bring it back to his home planet for study. Bullets did not harm Groot, and his wooden hide was "too tough to
burn." Groot's ability to mentally command other trees to move
and do his bidding quickly disabled the town's defenses, and all
seemed lost until the timely intervention of Leslie Evans, scientist.
Working nonstop for several days, Evans developed the one weapon
capable of immobilizing Groot—mutated termites. As shown in Figure 52, when the town's sheriff is chagrined that he "never even
thought of that," a relieved villager points out, "That's why Evans
is a scientist—and you're only a sheriff!" Meanwhile, Evans's wife
hugs her husband, declaring, "Oh, darling, forgive me! I've been
such a fool! I'll never complain about you again! Never!!" Person-
ally, I can't tell you how many times I've heard those very same
words from my own wife!†

While the scientist as world-saving hero is a caricature, I hope
that I have convinced you that the scientist as world-*changing* hero
is a pretty apt description for the physicists who developed the
field of quantum mechanics. In this, these investigators followed a
trail blazed hundreds of years ago. For science has *always* changed
the future. Technological innovations, from movable type to steam
engines to wireless radio to laptop computers, have time and again
profoundly altered interactions among people, communities, and
nations.

Discoveries in one field of science enable breakthroughs in oth-

---

* I don't want to tell them their jobs, but if I were an astronomer, I'd keep my eye on
Planet X. I think it might be trouble.

† Primarily because it involves imaginary numbers!

Figure 52: The final panel from Tales to Astonish # 13, showing Evans's reward for challenging Groot—the beginning of a "new, and better life" in which his wife "would never complain about [him] again!"

ers. The elucidation of the structure of DNA resulted from the interpretation of X-ray scattering data. This technique of X-ray spectroscopy was developed, through the application of quantum mechanics, to facilitate the study of crystalline structures by solid-state physicists. The deciphering of the human genome is inconceivable without the use of high-speed computers and data storage that rely on the transistor, invented over fifty years ago by scientists at Bell Labs. Using the tools developed by physicists in the last century, biologists in this century are poised to enact their own scientific revolution. Time will tell whether years from now another book will describe how "biologists changed the future." But one thing is for sure—we will not be able to embrace and participate in that future without the discipline, curiosity, questioning, and reasoning that science requires. And if Orrgo the Unconquerable (*Strange Tales* # 90) ever returns, we'll be ready!

ACKNOWLEDGMENTS

Sometimes, as the saying goes, the very best plan is to be lucky. I have been fortunate to have excellent professors when learning quantum mechanics, statistical mechanics and solid-state physics in college and graduate school. The first course I had in quantum physics was taught by Prof. Timothy Boyer, whose classroom instruction provided an excellent foundation in the topic while his research in developing a non-quantum explanation for atomic behavior (involving classical electrodynamics coupled with a zero-point radiation field) demonstrated that there was more than one way to view and account for natural phenomena. I am also happy to thank Herman Cummins, Fred W. Smith, Kenneth Rubin, William Miller, Robert Alfano, Robert Sachs, Leo Kadanoff, Hellmut Fritzsche, Sidney Nagel, and Robert Street, who taught me, in the classroom and out, about this fascinating field of physics and its many applications. My students at the University of Minnesota have been the inspiration and motivation for many of the examples presented here.

Writing a popular science book about quantum mechanics has been a challenging exercise, as it is very easy to trip up and misrepresent essential aspects of the theory in attempting to simplify the material for the non-expert. I am deeply grateful for the efforts of Benjamin Bayman, who read the entire manuscript in draft form and provided valuable feedback and corrections. In addition, William Zimmermann, E. Dan Dahlberg, Michel Janssen, Bruce Hammer, and Marco Peloso read portions of the text and I thank them for their insights and suggestions, along with the helpful comments of Yong-Zhong Qian, Paul Crowell, John Broadhurst, Allen Goldman, and Roger Stuewer. I thank Bruce Hammer for the magnetic resonance image in Chapter 19. Any errors or confusing arguments that remain are solely my responsibility.

    I am also grateful to Gotham Books in general, and my editor Patrick Mulligan in particular, for the opportunity to share with you the cool and practical field of quantum mechanics and its applications in nuclear and solid-state physics. Patrick's guidance during the writing and editing of this book has yielded a dramatically improved text, and his support from the very beginning in this book made it possible. The contributions of the copyeditor, Eileen Chetti, did much to improve the readability of the final manuscript. Michael Koelsch has done a wonderful job on the cover illustration and Elke Sigal on the book design. Christopher Jones did a fantastic job on the line drawings, beautifully illustrating complex ideas throughout the book. Thanks also to Alex Schumann and Brian Andersson for the electron and laser diffraction photos, and the pencil-in-glass shot (with thanks to Eric Matthies for the Jon Osterman pencil). Some of the science fiction magazines cited here were procured from Kayo Books in San Francisco, California, a great resource for all things pulpy. Travers Johnson at Gotham and Jake Sugarman at William Morris Endeavor Entertainment were of great help throughout the difficult process of seeing the manuscript from rough first draft through to its final state. My agent, Jay Mandel, has always had my back, and his insights, advice and encouragement throughout this project have been crucial. He's been there from the start and every step of the way.

    This book could not have been written without the limitless support and patience of my wife, Therese; and children, Thomas, Laura, and David, who graciously gave up their time with me while I was writing this book. I am grateful to Carolyn and Doug Kohrs for their friendship and support long before and throughout the writing of two books, and to Camille and Geoff Nash, who have always been there through thick and thin.

    As I struggled with the early drafts of the manuscript, the editing advice, research and counsel of my son Thomas and wife, Therese, have been invaluable. I am proud and honored to thank them for their hard work and encouragement. I have been luckiest of all to benefit from my family's love and support. I know that the future will exceed the predictions of the sunniest, most optimistic science fiction, as long as I share it with them.

INTRODUCTION

xi   "well into the twenty-first century, we still await flying cars, jet packs": *Follies of Science: 20th Century Visions of Our Fantastic Future*, Eric Dregni and Jonathan Dregni (Speck Press, 2006).

xii   "consider the long-term data storage accomplished by the Sumerians": *Ancient Mesopotamia: Portrait of a Dead Civilization*, A. Leo Oppenheim (University of Chicago Press, 1964); *The Sumerians*, C. Leonard Woolley (W. W. Norton and Co., 1965).

xii   "In 1965 Gordon Moore noted": *The Chip: How Two Americans Invented the Microchip and Launched a Revolution*, T. R. Reid (Random House, 2001).

CHAPTER 1

2   *Amazing Stories: Cheap Thrills, The Amazing! Thrilling! Astounding! History of Pulp Fiction*, Ron Goulart (Hermes, 2007).

3   "at the German Physical Society, Max Planck": *Thirty Years That Shook Physics: The Story of Quantum Theory*, George Gamow (Dover, 1985).

3   "Buck Rogers first appeared in the science fiction pulp *Amazing Stories*": *Science Fiction of the 20th Century: An Illustrated History*, Frank M. Robinson (Collectors Press, 1999).

3   "or what publisher Hugo Gernsback called 'scientifiction'": *Alternate Worlds: The Illustrated History of Science Fiction*, James Gunn (Prentice-Hall, 1975).

3   "Given the amazing pace of scientific progress": See, for example, *The Victorian Internet: The Remarkable Story of the Telegraph and the Nineteenth Century's On-Line Pioneers*, Tom Standage (Berkley Books, 1998); *Electric Universe: How Electricity Switched on the Modern World*, David Bodanis (Three Rivers Press, 2005).

4   "a revolution in physics occurred": *Thirty Years That Shook Physics: The Story of Quantum Theory*, George Gamow (Dover, 1985).

5   "'It is a great source of satisfaction to us'": "The Rise of Scientification," Hugo Gernsback, *Amazing Stories Quarterly* 1, 2 (Experimenter Publishing, Spring 1928).

6   "As Edward O. Wilson once cautioned": "The Drive to Discovery," Edward O. Wilson, *American Scholar* (Autumn 1984).

6   "Jules Verne considered the most extraordinary voyage of all": *Paris in the Twentieth Century*, Jules Verne (Random House, 1996).

9   "In one participant's recollection, Bohr proposed a theoretical model": *Thirty Years That Shook Physics: The Story of Quantum Theory*, George Gamow (Dover, 1985).

11   "Faraday was the first to suggest that electric charges and magnetic materials": *Electric Universe: How Electricity Switched on the Modern World*, David Bodanis (Three Rivers Press, 2005).

CHAPTER 2

13   "The Skylark of Space," Edward Elmer Smith, with Lee Hawkins Garby (uncredited) (The Buffalo Book Co., 1946); first serialized in *Amazing Stories*, 1928.

15   "this theory predicted results that were nonsensical": *Quantum Physics of Atoms, Molecules, Solids, Nuclei, and Particles*, Robert Eisberg and Robert Resnick (John Wiley and Sons, 1974).

16   "Planck firmly believed that light was a continuous electromagnetic wave": *The Quantum World: Quantum Physics for Everyone*, Kenneth W. Ford (Harvard University Press, 2004).

16   "In the late 1800s, physicists had discovered that certain materi-
     als": *The Strange Story of the Quantum*, 2nd edition, Banesh Hoff-
     man (Dover, 1959).

16   (footnote) "George Gamow, brilliant physicist and famed practical
     joker": *Eurekas and Euphorias: The Oxford Book of Scientific An-
     ecdotes*, Walter Gratzer (Oxford University Press, 2002).

17   "Lenard was working at the University of Heidelberg": *Quantum
     Legacy: The Discovery That Changed Our Universe*, Barry Parker
     (Prometheus Books, 2002).

22   "When Einstein wrote his paper on the 'photoelectric' effect": *Ein-
     stein: His Life and Universe*, Walter Isaacson (Simon and Schuster,
     2007).

23   "Fortunately for Einstein": *Quantum Legacy: The Discovery That
     Changed Our Universe*, Barry Parker (Prometheus Books, 2002).

23   "Millikan was one of the most careful and gifted experimentalists
     of his day": Ibid.

25   "Technically, a photon is defined as": "Light Reconsidered," Ar-
     thur Zajone, in *OPN Trends*, supplement to *Optics and Photonic
     News* ed. by Chandrasekhar Roychoudhuri and Rajarshi Roy, 14,
     S-2 (2003).

26   "As Albert Einstein reflected, 'All the fifty years of conscious
     brooding'": "The first phase of the Bohr-Einstein dialogue," Martin
     J. Klein, *Historical Studies in the Physical Sciences* 2, 1 (1970).

CHAPTER 3

27   *Science Wonder Stories* (Stellar Publishing Company, Feb. 1930).

29   "Goddard was an early example of a prominent scientist": *Differ-
     ent Engines: How Science Drives Fiction and Fiction Drives Sci-
     ence*, Mark L. Brake and Neil Hook (Macmillan, 2008).

31   "In 1923, Prince Louis de Broglie": *The Story of Quantum Me-
     chanics*, Victor Guillemin (Charles Scribner's Sons, 1968) ; *Physics
     and Microphysics*, Louis de Broglie (Pantheon, 1955).

34   "These crystalline arrangements of atoms can be used as atomic-scale 'oil slicks'": *Men Who Made a New Physics*, Barbara Lovett Cline (University of Chicago Press, 1987).

CHAPTER 4

38   "Jerry Siegel and Joe Shuster eventually sold their story": *Men of Tomorrow: Geeks, Gangsters and the Birth of the Comic Book*, Gerald Jones (Basic Books, 2004).

38   "As Jules Feiffer argued": *The Great Comic Book Heroes*, Jules Feiffer (Bonanza Books, 1965).

39   "It was proposed in 1925 that every fundamental particle behaves as if it is a spinning top": *Quantum Physics of Atoms, Molecules, Solids, Nuclei, and Particles*, Robert Eisberg and Robert Resnick (John Wiley and Sons, 1974).

40–41 *"Air Wonder Stories"*: *Pulp Culture: The Art of Fiction Magazines*, Frank M. Robinson and Lawrence Davidson (Collectors Press, 1998).

42   "In Isaac Asimov's novel": *Fantastic Voyage II: Destination Brain*, Isaac Asimov (Doubleday, 1987).

43   (footnote) "When I described Asimov's suggestion in my 2005 book": *Civil War Files*, Mark O'English (Marvel Comics, Sept. 2006).

44   "Does the experimentally observed magnetic field of electrons": *Quantum Physics of Atoms, Molecules, Solids, Nuclei, and Particles*, Robert Eisberg and Robert Resnick (John Wiley and Sons, 1974).

45   "Two Dutch students in Leiden, Samuel Goudsmit and George Uhlenbeck, wrote a paper in 1925": Ibid.

45   "'young enough to be able to afford a stupidity'": "George Uhlenbeck and the Discovery of Electron Spin," Abraham Pais, *Physics Today* (December 1989); *The Conceptual Development of Quantum Mechanics*, Max Jammer (McGraw-Hill, 1966).

CHAPTER 5

51   "In 1958, Jonathan Osterman": *Watchmen,* written by Alan Moore
      and drawn by Dave Gibbons (DC Comics, 1986, 1987).

51   "In addition to electromagnetism, the intrinsic field must be
      comprised of forces": *The Quantum World: Quantum Physics for
      Everyone,* Kenneth W. Ford (Harvard University Press, 2004).

52   "The neutron, discovered in 1932": *The Atom and Its Nucleus,*
      George Gamow (Prentice Hall, 1961).

52   (footnote) "when Guido, a superstrong member of a team of super-
      powered mutants": *X-Factor* # 72, written by Peter David and
      drawn by Larry Stroman (Marvel Comics, Nov. 1991).

56   "At the start of the twentieth century, physicists debated": *Thirty
      Years That Shook Physics: The Story of Quantum Theory,* George
      Gamow (Dover, 1985).

57   (footnote) "to quote Krusty the Clown": *The Simpsons,* episode
      # 3F08, "Sideshow Bob's Last Gleaning," written by Spike Feren-
      sten and directed by Dominic Polcino (Nov. 1995).

57   "An important step in reconciling this puzzle was Niels Bohr's
      suggestion in 1913": *Niels Bohr: A Centenary Volume,* edited by
      A. P. French and P. J. Kennedy (Harvard University Press, 1985).

59   "This newly detected element was named helium": *Helium: Child
      of the Sun,* Clifford W. Seibel (University Press of Kansas, 1968);
      *Reading the Mind of God: In Search of the Principle of Universal-
      ity,* James Trefil (Charles Scribner's Sons, 1989).

60   *Strange Adventures* # 62, "The Fireproof Man," written by
      John Broome and drawn by Carmine Infantino (DC Comics, Nov.
      1955).

60   (footnote) *Seduction of the Innocent,* Frederic Wertham (Rinehart
      Press, 1953).

62   "Following a seminar presentation by Schrödinger of the 'stand-
      ing wave' model of electrons in an atom": *Schrödinger: Life and*

Thought, Walter Moore (Cambridge University Press, 1989); *Quantum Legacy: The Discovery That Changed Our Universe*, Barry Parker (Prometheus Books, 2002).

CHAPTER 6

65 (footnote) "Historians of science debate to this day the identity of the woman": *Schrödinger: Life and Thought*, Walter Moore (Cambridge University Press, 1989); *Faust in Copenhagen: A Struggle for the Soul of Physics*, Gino Segre (Viking, 2007).

68 "When Schrödinger first solved this equation for the hydrogen atom": *Men Who Made a New Physics*, Barbara Lovett Cline (University of Chicago Press, 1987).

71 "One intriguing consequence of the Schrödinger equation": *Quantum Physics of Atoms, Molecules, Solids, Nuclei, and Particles*, Robert Eisberg and Robert Resnick (John Wiley and Sons, 1974).

72 "Schrödinger's equation destroyed the notion": *Men Who Made a New Physics*, Barbara Lovett Cline (University of Chicago Press, 1987).

CHAPTER 7

74 "Heisenberg's struggle to envision what the electron was doing": *Quantum Legacy: The Discovery That Changed Our Universe*, Barry Parker (Prometheus Books, 2002).

75 "In 1925 Heisenberg exiled himself to the German island of Helgoland": Ibid.

79 "It is the standard deviations of these two bell-shaped curves": *The Quantum World: Quantum Physics for Everyone*, Kenneth W. Ford (Harvard University Press, 2004).

82 "explains why scientists are concerned about an increase in the average global temperature of a few degrees": *Potential Impacts of Climate Change in the United States*, CBO Paper (Congressional Budget Office, May 2009).

83 "Fresh snow reflects 80 to 90 percent of all sunlight": *Energy in Nature and Society*, Vaclav Smil (MIT Press, 2008).

CHAPTER 8

85   "Dr. Manhattan": *Watchmen*, written by Alan Moore and drawn by Dave Gibbons (DC Comics, 1986, 1987).

85   "The quantum mechanical wave function contains all the information": *Quantum Physics of Atoms, Molecules, Solids, Nuclei, and Particles*, Robert Eisberg and Robert Resnick (John Wiley and Sons, 1974).

88   (figure caption) Image adapted from *The Atom and Its Nucleus*, George Gamow (Prentice Hall, 1961).

89   "this 'leakage effect' is observed for electrons": *The Atom and Its Nucleus*, George Gamow (Prentice Hall, 1961); *Thirty Years That Shook Physics: The Story of Quantum Theory*, George Gamow (Dover, 1985).

90   "the Schrödinger equation is linear": *Quantum Physics of Atoms, Molecules, Solids, Nuclei, and Particles*, Robert Eisberg and Robert Resnick (John Wiley and Sons, 1974).

91   "'nothing ever ends'": *Watchmen* # 12, written by Alan Moore and drawn by Dave Gibbons (DC Comics, Oct. 1987).

91   "Dave Gibbons, the artist of *Watchmen*, once stated in a radio interview": Interview with Dave Gibbons and James Kakalios, Minnesota Public Radio, *Science Friday*, hosted by Ira Flatow, March 27, 2009.

92   "they emit electromagnetic radiation in the blue-ultraviolet portion of the spectrum": *Introduction to High Energy Physics*, 2nd edition, Donald H. Perkins (Addison-Wesley, 1982).

94   "then as the wave function contains *all* the information": *Quantum Physics of Atoms, Molecules, Solids, Nuclei, and Particles*, Robert Eisberg and Robert Resnick (John Wiley and Sons, 1974).

95–96 "One interpretation was provided by Hugh Everett III": *The Many Worlds Interpretation of Quantum Mechanics*, edited by Bryce S. DeWitt and Neill Graham (Princeton University Press, 1973); *The Many Worlds of Hugh Everett III*, Peter Byrne (Oxford University Press, 2010).

96   *Flatland*, Edwin Abbott (Seely and Co., 1884).

96   "The Fifth Dimensional Catapult," Murray Leinster, first published in *Astounding* (January 1931).

96   "Plattner's Story," H. G. Wells, from *Thirty Strange Stories* (Harper and Bros., 1897).

96   "The Remarkable Case of Davidson's Eyes," H. G. Wells, from *The Stolen Bacillus and Other Incidents* (Methuen, 1895).

CHAPTER 9

101   "joined by the refrain of 'tic, tic, tic'": "Tic, Tic, Tic," sung by Doris Day in *My Dream Is Yours* (Warner Bros., 1949), music by Harry Warren and lyrics by Ralph Blaine.

101   *The Atomic Kid* (Republic Pictures, 1954), written by Blake Edwards, Benedict Freedman, and John Fenton Murray and directed by Leslie H. Martinson.

101–102   *The Beast of Yucca Flats*, written and directed by Coleman Francis (Image Entertainment, 1961).

102   "the true effects of radiation exposure were publicly known": "Hiroshima," John Hersey, *The New Yorker*, August 31, 1946, reprinted in *Reporting World War II: American Journalism 1938–1946* (Library of America, 1995).

102   "The 1957 television program *Disneyland* featured Dr. Heinz Haber": *The Walt Disney Story of Our Friend the Atom*, Heinz Haber (Dell Publishing, 1956).

102   *Across the Space Frontier*, edited by Cornelius Ryan, written by Werhner Von Braun, Oscar Schachter, and Willy Ley and illustrated by Chesley Bonestell (Viking, 1952).

103   "In 1957 Ford proposed a car called the Nucleon,": *Alien Hand Syndrome*, Alan Bellows (Workman Publishing Co., 2009).

103   *Twenty Thousand Leagues Under the Sea*, Jules Verne (Dodo Press, 2007).

104   (footnote) "The sponsor for the Disneyland program": *Fast Food Nation*, Eric Schlosser (Houghton Mifflin, 2004).

104   "When Erenst Rutherford's lab conducted experiments": *The Atom and Its Nucleus*, George Gamow (Prentice Hall, 1961); *Quantum Legacy: The Discovery That Changed Our Universe*, Barry Parker (Promethcus Books, 2002).

105   "For a while, physicists thought that the nucleus contained both protons and electrons": *Quantum Physics of Atoms, Molecules, Solids, Nuclei, and Particles*, Robert Eisberg and Robert Resnick (John Wiley and Sons, 1974).

106   (footnote) "To maintain stability in a nucleus requires a critical balance of the number of neutrons and protons": Ibid.

107   "A dictionary from the end of the nincteenth century": *The Walt Disney Story of Our Friend the Atom*, Heinz Haber (Dell Publishing, 1956).

107   "In 1937 Italian physicist Enrico Fermi": *The Atom and Its Nucleus*, George Gamow (Prentice Hall, 1961).

110   "Gilbert's U-238 Atomic Energy Lab": created by Alfred Carlton Gilbert (founder of the Gilbert Hall of Science in New York City in 1941), with consultation with the physics faculty at MIT (1951).

110   *Learn How Dagwood Splits the Atom*, written by John Dunning and Louis Hcil and drawn by Joe Musial (King Features Syndicate, 1949).

110   "Buck Rogers newspaper strip published in 1929": *Buck Rogers in the 25th Century: The Complete Newspaper Dailies*, vol. 1, *1929–1930*, written by Philip Nowlan and drawn by Richard Calkins (Hermes Press, 2008).

111   *Secret Service Operator No. 5*, issue # 47, written by Wayne Rogers (under the pen name Curtis Steele) (Sept. 1939).

111   *The World Set Free*, H. G. Wells (Quiet Vision Publishing, 2000).

112   *The Interpretation of Radium: Being the Substance of Six Free*

Popular Experimental Lectures Delivered at the University of Glasgow, Frederick Soddy (G. P. Putnam's Sons, 1909).

112 "This novel made a strong impression on one particular reader": *Different Engines: How Science Drives Fiction and Fiction Drives Science,* Mark L. Brake and Neil Hook (Macmillan, 2008).

CHAPTER 10

114 *Them!* written by Ted Sherdeman, Russell Hughes, and George Worthing Yates and directed by Gordon Douglas (Warner Bros., 1954).

114 *The Amazing Colossal Man,* written by Mark Hanna and Bert I. Gordon and directed by Bert I. Gordon (Malibu Productions, 1957).

114 *War of the Colossal Beast,* written by Bert I. Gordon and George Worthing Yates and directed by Bert I. Gordon (Carmel Productions,1958).

114 *The Beginning of the End,* written by Fred Freiberger and Lester Gorn and directed by Bert I. Gordon (AB-PT Pictures Corp., 1957).

114 *It Came from Beneath the Sea,* written by George Worthing Yates and Hal Smith and directed by Robert Gordon (Clover Productions, 1955).

114 *The Attack of the Giant Leeches,* written by Leo Gordon and directed by Bernard L. Kowalski (American International Pictures, 1959).

114 *The Incredible Shrinking Man,* written by Richard Matheson and directed by Jack Arnold (Universal, 1957).

114 *Dr. Cyclops,* written by Tom Kilpatrick and directed by Ernest B. Schoedsack (Paramount, 1940).

116 "Even though the nucleus can lower its energy by ejecting an alpha particle": *The Atom and Its Nucleus,* George Gamow (Prentice Hall, 1961).

117 "One important similarity between the lottery scenario and the decay of unstable nuclei": *The Quantum World: Quantum Physics for Everyone*, Kenneth W. Ford (Harvard University Press, 2004).

119 *Learn How Dagwood Splits the Atom*, written by John Dunning and Louis Heil and drawn by Joe Musial (King Features Syndicate, 1949).

119–120 "why is there any tritium still around": *The Atom and Its Nucleus*, George Gamow (Prentice Hall, 1961).

120 "Sheldon Kauffman and Willard F. Libby did the next best thing": "The Natural Distribution of Tritium," Sheldon Kaufman and W. F. Libby, *Physical Review* 93 (1954), p. 1337. See http://link.aps.org/abstract/PR/v93/p1337 for this article.

124 "Elements that emit gamma rays, alpha particles, or beta particles": *Quantum Physics of Atoms, Molecules, Solids, Nuclei, and Particles*, Robert Eisberg and Robert Resnick (John Wiley and Sons, 1974).

125 (figure caption) Buck Rogers ray-gun image from *Raygun*, Eugene W. Metcalf and Frank Maresca, photographed by Charles Bechtold (Fotofolio, 1999).

125 (figure caption) Penetration of matter by radiation image adapted from www.hyperphysics.phys-astr.gsu.edu Web site.

126 "Russian journalist Alexander Litvinenko was murdered": *The Terminal Spy: A True Story of Espionage, Betrayal and Murder*, Alan S. Crowell (Broadway, 2008).

127 "When some of these particles strike the DNA in our cells": "Cosmic Rays: A Review for Astrobiologists," Franco Ferrari and Ewa Szuszkiewicz, *Astrobiology* 9, 413 (2009).

CHAPTER 11

128 "In the 1962 Gold Key comic book series": *Doctor Solar—Man of the Atom*, written by Paul S. Newman and drawn by Bob Fujitani (Gold Key Comics, 1962), reprinted in *Doctor Solar—Man of the Atom*, vols. 1–4 (Dark Horse Books, 2004–2008).

129   "Captain Atom": *Space Adventures* # 33, written by Joe Gill and Steve Ditko and drawn by Steve Ditko (Charlton Comics, Mar. 1960), reprinted in *Action Hero Archives*, vol. 1 (DC Comics, 2004).

129   *Dr. Solar, Man of the Atom* # 14, written by Paul S. Newman and drawn by Frank Bolle (Gold Key Comics, Sept. 1965), reprinted in *Doctor Solar—Man of the Atom*, vol. 2 (Dark Horse Books, 2005).

130   "Neutrons themselves are not stable": *The Story of Quantum Mechanics*, Victor Guillemin (Charles Scribner's Sons, 1968).

131   "When physicists in the late 1920s discovered this phenomenon": *Thirty Years That Shook Physics: The Story of Quantum Theory*, George Gamow (Dover, 1985).

132   "fortunately this inverse process occurs constantly in the center of the sun": "Nuclear Astrophysics," M. Arnould and K. Takahashi, *Rep. Prog. Phys.* 62, 395 (1999).

133   (figure caption) Image adapted from www.hyperphysics.phys-astr .gsu.edu Web site.

134   (footnote) "To this Eddington replied in 1920": "Arthur Stanley Eddington: A Centennial Tribute," Joe S. Tenn, *Mercury* 11 (1982), p. 178.

135   "the average photon spends forty thousand years colliding": "How Long Does It Take for Heat to Flow Through the Sun?" G. Fiorentini and B. Rici, *Comments on Modern Physics* 1 (1999), p. 49.

135   "Our sun converts a great deal of hydrogen": "The Evolution and Explosion of Massive Stars," S. E. Woolsey and A. Heger, *Reviews of Modern Physics* 74 (Oct. 2002), p. 1015.

136   "For more than fifty years, scientists have been attempting to construct a fusion reactor": *Sun in a Bottle*, Charles Seife (Viking, 2008); "Fusion's False Dawn," Michael Moyer, *Scientific American* 302, 50 (2010).

136   "This so-called cold fusion process": *Bad Science: The Short Life and Weird Times of Cold Fusion*, Gary Taubes (Random House, 1993).

CHAPTER 12

141 "The agreement between theoretical predictions": *QED: The Strange Theory of Light and Matter*, Richard P. Feynman (Princeton University Press, 1988).

142 "The collective behavior of quantum particles": *Quantum Physics of Atoms, Molecules, Solids, Nuclei, and Particles*, Robert Eisberg and Robert Resnick (John Wiley and Sons, 1974).

144 "Think about a ribbon": The argument involving a ribbon with different colors on each side is adapted from that in "The Reason for Antiparticles," Richard P. Feynman, in *Elementary Particles and the Laws of Physics*, Richard P. Feynman and Steven Weinberg (Cambridge University Press, 1987). Feynman credits this analogy to David Finkelstein.

146 "described as the difference of two functions, A and B": *The Quantum World: Quantum Physics for Everyone*, Kenneth W. Ford (Harvard University Press, 2004).

148 "Wolfgang Pauli, one of the founding fathers of quantum mechanics": *Thirty Years That Shook Physics: The Story of Quantum Theory*, George Gamow (Dover, 1985).

150 "One easy way to satisfy the Pauli principle": *Electrons in Metals: A Short Guide to the Fermi Surface*, J. M. Ziman (Taylor and Francis, 1963).

CHAPTER 13

155 *The Superorganism—The Beauty, Elegance, and Strangeness of Insect Societies*, Bert Hölldobler and Edward O. Wilson (W. W. Norton & Co., 2008).

155 *The Cosmic Rape*, Theodore Sturgeon (Pocket Books, 1958).

155 (footnote) *More Than Human*, Theodore Sturgeon (Farrar, Straus and Young, 1953).

156 "Bose-Einstein condensate": *The Quantum World: Quantum Physics for Everyone*, Kenneth W. Ford (Harvard University Press, 2004).

158   "This symmetry indicates that the two-particle wave function": Ibid.

160   *Village of the Damned*, written by Stirling Silliphant, Wolf Rilla, and Ronald Kinnoch and directed by Wolf Rilla (Metro-Goldwyn-Mayer, 1960).

161   "helium is an example of an atom": *Quantum Physics of Atoms, Molecules, Solids, Nuclei, and Particles*, Robert Eisberg and Robert Resnick (John Wiley and Sons, 1974).

162   "Experimentalists in 1965," "Gyroscopic Detection of Persistent Flow of Super fluid Liquid Helium," J.B. Mehl and W. Zimmermann Jr., *Physical Review Letters* 14, p. 815 (1965).

163   "an electrical analog to superfluidity": *The Path of No Resistance: The Story of a Revolution in Superconductivity*, Bruce Schechter (Simon and Schuster, 1989); Introduction to Superconductivity, 2nd Edition, Michael Tinkham (Dover, 2004); "Observation of Persistent Current in a Superconducting Solenoid," J. File and R.G. Mills, *Physical Review of Letters* 10, p. 93 (1963).

166   "Max Planck and how his explanation of the spectrum of light": *Quantum Legacy: The Discovery That Changed Our Universe*, Barry Parker (Prometheus Books, 2002); *Men Who Made a New Physics*, Barbara Lovett Cline (University of Chicago Press, 1987).

167   "What if the box were filled with light": *Thirty Years That Shook Physics: The Story of Quantum Theory*, George Gamow (Dover, 1985).

168   "The box containing light can be considered a gas of photons": Ibid.; *The Atom and Its Nucleus*, George Gamow (Prentice Hall, 1961).

CHAPTER 14

173   "In our world the eighty-sixth floor of the Empire State Building": *Doc Savage: His Apocalyptic Life*, Philip Joes Farmer (Doubleday, 1973); *A History of the Doc Savage Adventures*, Robert Michael "Bobb" Cotter (McFarland and Company, 2009).

174   "Dent and his wife lived for several years on a forty-foot schooner": *Lester Dent: The Man, His Craft and His Market*, M. Martin

McCarey-Laird (Hidalgo Pub. Co., 1994); *Bigger Than Life: The Creator of Doc Savage*, Marilyn Cannaday (Bowling Green State University Popular Press, 1990).

174 "called his Fortress of Solitude": *Fortress of Solitude*, Lester Dent (under the pen name Kenneth Robeson) (Street and Smith, 1938); reprinted in *Doc Savage* # 1 (Nostalgia Ventures, 2006).

174 "his tie and jacket buttons hid . . . and his car could produce": "The Bronze Genius," Will Murray in *The Man Behind Doc Savage: A Tribute to Lester Dent*, edited by Robert Weinberg (Weinberg, 1974).

174 "the comic-book superhero foursome": "Introduction to *The Fortress of Solitude*" in *Doc Savage* # 1 (Nostalgia Ventures, 2006).

174 "Doc's gadgets were similarly ahead of his time": "The Bronze Genius," Will Murray in *The Man Behind Doc Savage: A Tribute to Lester Dent*, edited by Robert Weinberg (Weinberg, 1974).

174 "According to Dent, a reference to radar": *Doc Savage: His Apocalyptic Life*, Philip José Farmer (Doubleday, 1973).

175 *The Man of Bronze*, Lester Dent (under the pen name Kenneth Robeson) (Street and Smith, 1933); reprinted in *Doc Savage* # 14 (Nostalgia Ventures, 2008).

175 (footnote) "'Ralph 124C 41+'": The relevant paragraph, containing a detailed description of what we would today term "radar," is reproduced in *Alternate Worlds: The Illustrated History of Science Fiction*, James Gunn (Prentice-Hall, 1975).

177 "What determines these transition rates": *Quantum Physics of Atoms, Molecules, Solids, Nuclei, and Particles*, Robert Eisberg and Robert Resnick (John Wiley and Sons, 1974); *QED: The Strange Theory of Light and Matter*, Richard P. Feynman (Princeton University Press, 1988).

CHAPTER 15

183 "the demand for Buck Rogers– and Flash Gordon–inspired toy ray guns": *Raygun*, Eugene W. Metcalf and Frank Maresca, photographed by Charles Bechtold (Fotofolio, 1999).

183   "At his press conference in 1960, Maiman was peppered": *The Laser Odyssey*, T. H. Maiman (Laser Press, 2000).

183   "scientists from Bell Labs were instructed by management": "Bell Labs and the Ruby Laser," D. F. Nelson, R. J. Collins, W. Kaiser, *Physics Today* 63, 40 (2010).

184   *Goldfinger*, written by Richard Maibaum and Paul Dehn and directed by Guy Hamilton (Metro-Goldwyn-Mayer, 1964).

184   "The circular buzz saw of the 1959 novel": *Goldfinger*, Ian Fleming (Jonathan Cape, 1959).

184   "How can one ensure that all the electrons residing in the laser levels": *Lasers and Holography: An Introduction to Coherent Optics*, 2nd edition, Winston Kock (Dover, 1981).

184   (footnote) "a mixture of two gases, helium and neon": *Quantum Legacy: The Discovery That Changed Our Universe*, Barry Parker (Prometheus Books, 2002).

185   (footnote) "the above argument applies to electric dipole transitions": *Quantum Physics of Atoms, Molecules, Solids, Nuclei, and Particles*, Robert Eisberg and Robert Resnick (John Wiley and Sons, 1974); *QED: The Strange Theory of Light and Matter*, Richard P. Feynman (Princeton University Press, 1988).

186   "The device produces *l*ight *a*mplification by *s*timulated *e*mission of *r*adiation": *Lasers and Holography: An Introduction to Coherent Optics*, 2nd edition, Winston Kock (Dover, 1981).

187   "sent out from a lab on Earth": *They All Laughed: From Light Bulbs to Lasers: The Fascinating Stories Behind the Great Inventions That Have Changed Our Lives*, Ira Flatow (Harper Perennial, 1992).

187   "Both of these elements . . . are thus chemically inert": *Quantum Physics of Atoms, Molecules, Solids, Nuclei, and Particles*, Robert Eisberg and Robert Resnick (John Wiley and Sons, 1974).

188   "Digital video discs (DVDs) and compact discs (CDs)": *How Everything Works: Making Physics Out of the Ordinary*, Louis A. Bloomfield (John Wiley and Sons, 2008).

190   "Birth of the Death Ray,": *Dr. Solar—Man of the Atom* # 16, written by Paul S. Newman and drawn by Frank Bolle (Gold Key Com-

ics, June 1966), reprinted in *Dr. Solar—Man of the Atom*, vol. 2 (Dark Horse Books, 2005).

191   "the number of Gillette razor blades they could melt through": *They All Laughed: From Light Bulbs to Lasers: The Fascinating Stories Behind the Great Inventions That Have Changed Our Lives*, Ira Flatow (Harper Perennial, 1992).

192   "Superman removes the thick cataracts that have blinded a companion": "A Matter of Light and Death," *Action* # 491, written by Cary Bates and drawn by Curt Swan (DC Comics, Jan. 1979).

192   "Eight years later, Dr. Stephen Trokel": *Excimer Lasers in Ophthalmology*, edited by David S. Gartry (Informa Healthcare, 1997).

193   *The South Pole Terror*, Lester Dent (under the pen name Kenneth Robeson) (Street and Smith, 1936); reprinted in *Doc Savage* # 11 (Nostalgia Ventures, 2007).

CHAPTER 16

194   "robots would break free": *Follies of Science: 20th Century Visions of Our Fantastic Future*, Eric Dregni and Jonathan Dregni (Speck Press, 2006).

194   "The Challengers were four adventurers": *Showcase* # 6, written by Dave Wood and drawn by Jack Kirby (DC Comics, Jan.–Feb. 1957); reprinted in *Challengers of the Unknown Archives*, vol. 1 (DC Comics, 2003).

195   "The Challengers' challenge": "ULTIVAC is Loose!," *Showcase* # 7, written by Dave Wood and drawn by Jack Kirby (DC Comics, Mar.–Apr. 1957); reprinted in *Challengers of the Unknown Archives*, vol. 1 (DC Comics, 2003).

197   "In 1946, scientists at the University of Pennsylvania": *Quantum Legacy: The Discovery That Changed Our Universe*, Barry Parker (Prometheus Books, 2002).

197   "a Bell Labs scientist, Russell Ohl": *Crystal Fire: The Birth of the Information Age*, Michael Riordan and Lillian Hoddeson (W. W. Norton and Co., 1997).

199 "Semiconductors make convenient light detectors": Ibid.

200 "smoke detector": *The Way Things Work*, David Macaulay (Houghton Mifflin, 1988); *How Everything Works: Making Physics Out of the Ordinary*, Louis A. Bloomfield (John Wiley and Sons, 2008).

201 "The Shadow, who in reality is Lamont Cranston, wealthy man-about-town": *The Shadow Scrapbook*, Walter B. Gibson (Harcourt Brace Jovanovich, 1979).

201 *Death Stalks the Shadow*, author unknown (original air date October 9, 1938, on the Mutual Network).

202 "when different chemical impurities are added": *Introduction to Solid State Physics*, 7th edition, Charles Kittel (John Wiley and Sons, 1995).

205 "the junction between the p-type and n-type": *Crystal Fire: The Birth of the Information Age*, Michael Riordan and Lillian Hoddeson (W. W. Norton and Co., 1997).

205 "Solid-state semiconductor diodes": *Physics of Semiconductor Devices*, 2nd edition, S. Sze (Wiley-Interscience, 1981); *Electronics: Circuits and Devices*, 2nd edition, Ralph J. Smith (John Wiley and Sons, 1980).

206 two regions of an identical semiconductor are separated by a very thin insulator": Ibid.

206 "In 1939, Russell Ohl, a scientist at Bell Labs, was studying the electrical properties of semiconductors": *Crystal Fire: The Birth of the Information Age*, Michael Riordan and Lillian Hoddeson (W. W. Norton and Co., 1997).

207 "a light-emitting diode (LED) is a pn junction": *Physics of Semiconductor Devices*, 2nd edition, S. Sze (Wiley-Interscience, 1981); Electronics: Circuits and Devices, 2nd edition, Ralph J. Smith (John Wiley and Sons, 1980).

208 "In the past thirty years the luminosity of these devices": "From Transistors to Lasers to Light-Emitting Diodes," Nick Holonyak

Jr., *MRS Bulletin*, 30, 509 (2005); "The Quest for White LEDs Hits the Home Stretch," Robert F. Service, *Science* 325, 809 (2009).

CHAPTER 17

211 "the invention of the transistor": *The Chip: How Two Americans Invented the Microchip and Launched a Revolution*, T. R. Reid (Random House, 2001).

211 "the field-effect structure": *Crystal Fire: The Birth of the Information Age*, Michael Riordan and Lillian Hoddeson (W. W. Norton and Co., 1997).

214 "In 1958, just a year after the Challengers". Ibid.

215 "Estimates of the number of transistors": *Astronomy: The Solar System and Beyond*, 2nd edition, Michael A. Seeds (Brooks/Cole, 2001); "One Billion Transistors, One Uniprocessor, One Chip," Yale N. Patt, Sanjay Patel, Marius Evers, Daniel H. Friendly, and Jared Stark, *IEEE Computer* (September 1997).

215 "the transport of a single electron can be detected": "The Single Electron Transistor," M. A. Kastner, *Rev. Mod. Phys.*, 64, 849 (1992); "Artificial Atoms," Marc A. Kastner, *Physics Today* (January 1993).

215 "a complex literature using an alphabet consisting of only two letters": *The Chip: How Two Americans Invented the Microchip and Launched a Revolution*, T. R. Reid (Random House, 2001).

217 "Flash memory devices add a second metal electrode": "A Floating-Gate and Its Application to Memory Devices," D. Kahng and S. M. Sze, *The Bell System Technical Journal* 46 (1967), p. 1288; "Introduction to Flash Memory," R. Bez, E. Camerlenghi, A. Modelli, and A. Visconti, *Proceedings of the IEEE* 91 (2003), p. 489; "Reviews and Prospects of Non-Volatile Semiconductor Memories," F. Masuoka, R. Shirota, and K. Sakui, *IEICE Transactions* E 74 (1991), p. 868.

217 "What's the point of the second metal layer": "Volatile and Non-Volatile Memories in Silicon with Nano-Crystal Storage," S. Ti-

wari, F. Rana, K. Chan, H. Hanafi, W. Chan, and D. Buchanan, *IEDM Technical Digest* (1995), p. 521.

219   *Amazing Stories* (Teck Publications, December 1936).

219–220   "Wrist phones that are capable of video transmission": "Bell Labs Reports Progress on 'Dick Tracy' Watch," *APS News* 8, 6 (June 1999); "Yesterday's Dreams and Today's Reality in Telecommunications," Robert W. Lucky, *Technology in Society* 26 (2004), p. 223.

CHAPTER 18

221   "devices characterized as 'spintronic'": "Spintronics," David D. Awschalom, Michael E. Flatte, and Nitin Samarth, *Scientific American* (June 2002).

222   "'giant magnetoresistance'": "Giant Magnetoresistance of (001)Fe/ (001)Cr Magnetic Superlattices," M. N. Baibich , J. M. Broto, A. Fert, F. Nguyen Van Dau, F. Petroff, P. Eitenne, G. Creuzet, A. Friederich, and J. Chazelas, *Physical Review Letters* 61 (1988), p. 2472; "Enhanced Magnetoresistance in Layered Magnetic Structures with Antiferromagnetic Interlayer Exchange," G. Binasch, P. Grünberg, F. Saurenbach, and W. Zinn, *Physical Review B* 39 (1989), p. 4828.

223   "In any real metal wire there will be defects": *Introduction to Solid State Physics*, 7th edition, Charles Kittel (John Wiley and Sons, 1995).

223–224   "Imagine a flow of electrons perpendicular": "Spintronics," David D. Awschalom, Michael E. Flatte, and Nitin Samarth, *Scientific American* (June 2002).

225   "magnetic sensors on hard drives that employ another quantum mechanical phenomenon—tunneling": "Frontiers in spin-polarized tunneling," Jagadeesh S. Moodera, Guo-Xing Miao, and Tiffany S. Santos, *Physics Today* (April 2010).

226   "the first (expensive) transistor radio": *The Chip: How Two Americans Invented the Microchip and Launched a Revolution*, T. R. Reid (Random House, 2001).

CHAPTER 19

227 *X: The Man with the X-ray Eyes,* written by Robert Dillon and Ray Russell and directed by Roger Corman (Alta Vista Productions, 1963).

228 "Associated with the spin is a small intrinsic magnetic field": *Quantum Physics of Atoms, Molecules, Solids, Nuclei, and Particles,* Robert Eisberg and Robert Resnick (John Wiley and Sons, 1974).

230 "The idea begins to form": *How Does MRI Work? An Introduction to the Physics and Function of Magnetic Resonance Imaging,* Dominik Weishaupt, Victor D. Koechli, and Borut Marincek (Springer, 2008).

230 "spatial resolution throughout a cross section": *Naked to the Bone: Medical Imaging in the Twentieth Century,* Bettyann Holtzmann Kevles (Rutgers University Press, 1997).

233 "functional magnetic resonance imaging": *Introduction to Functional Magnetic Resonance Imaging: Principles and Techniques,* Richard B. Buxton (Cambridge University Press, 2002).

234 *The Demolished Man,* Alfred Bester (Shasta Publishers, 1953).

234 *More Than Human,* Theodore Sturgeon (Farrar, Straus and Young, 1953).

234 *The Cosmic Rape,* Theodore Sturgeon (Pocket Books, 1958).

234 *Village of the Damned,* written by Stirling Silliphant, Wolf Rilla, and Ronald Kinnoch and directed by Wolf Rilla (Metro-Goldwyn-Mayer, 1960).

235 "to directly discern a person's thoughts and intentions": "The Brain on the Stand: How Neuroscience Is Transforming the Legal System," Jeffrey Rosen, *The New York Times Magazine* (March 11, 2007); "Duped," Margaret Talbot, *The New Yorker* (July 2, 2007); "Head Case," Virginia Hughes, *Nature* 464, 340 (March 18, 2010).

CHAPTER 20

239   "In this way the 'ones' and 'zeros'": *The Chip: How Two Ameri-cans Invented the Microchip and Launched a Revolution*, T. R. Reid (Random House, 2001).

240   "analog-to-digital converter": *How Everything Works: Making Physics Out of the Ordinary*, Louis A. Bloomfield (John Wiley and Sons, 2008).

240   "the newest multitouch versions": "Hands-On Computing," Stuart F. Brown, *Scientific American* (July 2008).

241   "'semiconductor spintronic' devices": "Spintronics," David D. Awschalom, Michael E. Flatte, and Nitin Samarth, *Scientific American* (June 2002).

242   "resulting temperature rise . . . can limit the integrated circuit's performance": Electronics: Circuits and Devices, 2nd edition, Ralph J. Smith (John Wiley and Sons, 1980).

242   "A 'quantum computer' is a different beast entirely": *A Shortcut Through Time: The Path to the Quantum Computer*, George Johnson (Vintage Books, 2003).

244   "A small-scale prototype quantum computer": "Algorithms for Quantum Computation: Discrete Logarithms and Factoring," Peter Shor, *Proceedings of the 35th Annual Symposium on Foundations of Computer Science*, p. 124 (IEEE Computer Society Press, 1994); an accessible summary of Shor's algorithm can be found in Chapter 5 of *A Shortcut Through Time: The Path to the Quantum Computer*, George Johnson (Vintage Books, 2003).

244   *Transformers*, written by Roberto Orci and Alex Kurtzman and directed by Michael Bay (Dreamworks, 2007).

244   "the reason that the quantum ribbon can represent all four possible outcomes simultaneously": *Teleportation: The Impossible Leap*, David Darling (John Wiley and Sons, 2005).

245   "Einstein smelled a rat in this scenario": Ibid.; *The God Effect: Quantum Entanglement, Science's Strangest Phenomenon*, Brian Clegg (St. Martin's Griffin, 2006).

246 "'spooky action at a distance'": *Einstein: His Life and Universe,*
Walter Isaacson (Simon and Schuster, 2007).

246 "Books have been written over the question": And I'm not kidding!
See, for example, *The Physics of Quantum Information: Quantum
Cryptography, Quantum Teleportation, Quantum Computation,*
edited by D. Bouwmeester, A. K. Ekert, and A. Zeilinger (Springer-
Verlag, 2000); *Entanglement: The Greatest Mystery in Physics,*
Amir D. Aczel (John Wiley and Sons, 2002); *A Shortcut Through
Time: The Path to the Quantum Computer,* George Johnson (Vin-
tage Books, 2003); *Quantum Computing,* 2nd edition, Mika
Hirvensalo (Springer-Verlag, 2004); *Teleportation: The Impossible
Leap,* David Darling (John Wiley and Sons, 2005); *The God Effect:
Quantum Entanglement, Science's Strangest Phenomenon,* Brian
Clegg (St. Martin's Griffin, 2006); *The Age of Entanglement: When
Quantum Physics Was Reborn,* Louisa Gilder (Alfred A. Knopf,
2008).

246 "the two electrons' wave functions must remain 'entangled'": *A
Shortcut Through Time: The Path to the Quantum Computer,*
George Johnson (Vintage Books, 2003).

247 "recent experiments in 'teleportation'": *Teleportation: The Impossible Leap,* David Darling (John Wiley and Sons, 2005).

247 "experiments . . . . concerning two entangled quantum entities":
"Experimental Entanglement Swapping: Entangling Photons That
Never Interacted," Jian-Wei Pan, Dik Bouwmeester, Harald Wein-
furter, and Anton Zeilinger, *Physical Review Letters* 80, 3891
(1998), "Experiment and the Foundations of Quantum Physics,"
Anton Zeilinger, *Reviews of Modern Physics* 71 (1999), p. S288.

248 "from a 1998 issue of the adventures of the Justice League of America": *JLA* # 19, written by Mark Waid and drawn by Howard Porter
(DC Comics, 1998).

CHAPTER 21

249 "'cavorite'": Discovered by Dr. Cavor as described in *The First
Men in the Moon,* H. G. Wells (George Newnes, 1901).

249   "how much energy it takes to lift": *Conceptual Physics,* Paul G. Hewitt (Prentice Hall, 2002).

249   "every chemical reaction . . . on the order of an electron Volt": *Quantum Physics of Atoms, Molecules, Solids, Nuclei, and Particles,* Robert Eisberg and Robert Resnick (John Wiley and Sons, 1974).

250   "prototype jet packs have been able to keep test pilots aloft": *Jetpack Dreams: One Man's Up and Down (But Mostly Down) Search for the Greatest Invention That Never Was,* Mac Montandon (Da Capo Press, 2008).

250   *Iron Man,* written by Mark Fergus, Hawk Ostby, Art Marcum, and Matt Holloway and directed by Jon Favreau (Marvel Studios, 2008).

251   "According to the World Health Organization": "How Hard Is It to Convert Seawater to Fresh Drinking Water?" Ethan Trex, *Mental Floss* (August 2009).

251   "Global consumption of energy, which in 2005": *Energy,* Vaclav Smil (Oneworld Publications, 2006).

252   "The surface of the Earth receives": Ibid.

252   "present production capacity": "High Growth Reported for the Global Photovoltaic Industry," Reuters (Mar. 3, 2009); "A Solar Grand Plan," Ken Zweibel, James Mason, and Vasilis Fthenakis, *Scientific American* (Jan. 2008).

253   "two scientists, Johannes Bednorz and Karl Müller": *The Path of No Resistance: The Story of the Revolution in Superconductivity,* Bruce Schechter (Simon and Schuster, 1989).

256   "'thermoelectrics'": *Thermoelectrics Handbook: Macro to Nano,* edited by D. M. Rowe (CRC, 2005).

257   "extract electrical power from random vibrations involves nanogenerators": "Self-Powered Nanotech," Zhong Lin Wang, *Scientific American* (January 2008); "Nanogenerators Tap Waste Energy to Power Ultrasmall Electronics," Robert F. Service, *Science* 328, 304 (2010).

258 "In a battery, making use of essentially a reverse electrolysis process": *Batteries in a Portable World: A Handbook on Rechargeable Batteries for Non-Engineers*, 2nd Edition, Isidor Buchmann (Cadex Electronics, 2001); *The Battery: How Portable Power Sparked a Technological Revolution*, Henry Schlesinger (Smithsonian, 2010).

259 "improvements in the energy content and storage capacity of rechargeable batteries": *Batteries in a Portable World: A Handbook on Rechargeable Batteries for Non-Engineers*, 2nd Edition, Isidor Buchmann (Cadex Electronics, 2001)

260 "Nanotextured electrodes": "Nanostructured Electrodes and the Low-Temperature Performance of Li-Ion Batteries," Charles R. Sides and Charles R. Martin, *Advanced Materials*, 17, 128 (2005); "High-Rate, Long-Life Ni-Sn Nanostructured Electrodes for Lithium-Ion Batteries," J. Hassoun, S. Panero, P. Simon, P. L. Taberna, and B. Scrosati, *Advanced Materials*, 19, 1632 (2007).

260 "silicon nanoscale wires": "High-Performance Lithium Battery Anodes Using Silicon Nanowires," C. K. Chan, H. Peng, G. Liu, K. McIlwrath, X. F. Zhang, R. A. Huggins, and Y. Cui, *Nature Nanotechnology* 3, 31 (2008).

260 "Nanoscale filaments woven into textiles": "Smart Nanotextiles: A Review of Materials and Applications," S. Coyle, Y. Wu, K.-T. Lau, D. DeRossi, G. Wallace, and D. Diamond, *Materials Research Society Bulletin* 32 (May 2007), p. 434.

260 "highly refined pharmaceutical delivery systems": "Less Is More in Medicine," A. Paul Alivisatos, *Scientific American Reports* 17 (2007), p. 72.

AFTERWORD

262 "Buck Rogers newspaper strips": *Buck Rogers in the 25th Century: The Complete Newspaper Dailies*, vol. 1, *1929–1930*, written by Philip Nowlan and drawn by Richard Calkins (Hermes Press, 2008).

263 "New Wiring Idea May Make the All-Electric House Come True,": *Science Illustrated* (May 1949).

264   "in 1960 sales of *Superman* comics": *The Ten-Cent Plague: The Great Comic-Book Scare and How It Changed America*, David Hajdu (Farrar, Straus and Giroux, 2008).

264   *Adventure Comics* # 247, written by Otto Binder and drawn by Al Plastino (DC Comics, 1958); reprinted in *Legion of Superheroes Archives*, vol. 1 (DC Comics, reissue edition, 1991).

265   *Adventure* # 321, written by Edmond Hamilton and drawn by John Forte (DC Comics, June 1964); reprinted in *Showcase Presents Legion of Superheroes*, vol. 1 (DC Comics, 2007).

265–266   "Similarly, over at Marvel Comics": See, for example, *Marvel Masterworks Atlas Era Tales to Astonish*, vol. 1 (Marvel Comics, 2006) and vol. 2 (Marvel Comics, 2008); *Marvel Masterworks Atlas Era Tales of Suspense*, vol. 1 (Marvel Comics, 2006) and vol. 2 (Marvel Comics, 2008); *Amazing Fantasy Omnibus* (Marvel Comics, 2007).

267   *Tales to Astonish* #13, "I Challenged Groot! the Monster from Planet X!" written by Stan Lee and Larry Lieber and drawn by Jack Kirby (Marvel Comics, Nov. 1960); reprinted in *Marvel Masterworks Atlas Era Tales to Astonish*, vol. 2 (Marvel Comics, 2008).

268   *Strange Tales* # 90, "Orrgo . . . the Unconquerable," written by Stan Lee and Larry Lieber and drawn by Jack Kirby (Marvel Comics, Nov. 1961).

As mentioned in the introduction, one reason I have avoided an historical approach to relating the principles of quantum mechanics (aside from the not inconsequential fact that I am not an historian of science) is that there already exist many excellent histories of this period in physics. Readers interested in learning more about such questions as "what did Bohr know and when did he know it?" may enjoy *Thirty Years That Shook Physics: The Story of Quantum Theory* by George Gamow (Dover, 1985) as well as *The Great Physicists from Galileo to Einstein* (Dover, 1988) by the same author; Barbara Lovett Cline's *Men Who Made a New Physics* (University of Chicago Press, 1987); *Quantum Legacy: The Discovery that Changed Our Universe* by Barry Parker (Prometheus Books, 2002); *Reading the Mind of God: In Search of the Principles of Universality* by James Treffil (Anchor, 1989); Brian Cathcart's *The Fly in the Cathedral: How a Group of Cambridge Scientists Won the International Race to Split the Atom* (Farrar, Straus & Giroux, 2005); and Gino Segre's *Faust in Copenhagen: A Struggle for the Soul of Physics* (Viking, 2007).

The real superheroes of science who pioneered this field of physics receive their due in several excellent biographies, such as *Niels Bohr: A Centenary Volume*, edited by A. P. French and P. J. Kennedy (Harvard University Press, 1985); *The Strangest Man: The Hidden Life of Paul Dirac, Mystic of the Atom* by Graham Farmelo (Basic Books, 2009); *Beyond Uncertainty: Heisenberg, Quantum Physics and the Bomb*, David C. Cassidy (Bellevue Literary Press, reprinted in 2010); *Schrödinger: Life and Thought*, Walter Moore (Cambridge University Press, 1989); Abraham Pais's *Niels Bohr's Times: in Physics, Philosophy and Polity* (Claredon Press/Oxford

University Press, 1991) and *Subtle is the Lord: The Science and the Life of Albert Einstein* by (Oxford University Press, 1982); *Einstein: His Life and Universe,* Walter Isaacson (Simon and Schuster, 2007); Jeremy Bernstein's *Oppenheimer: Portrait of an Enigma* (Ivan R. Dee, 2004); *The End of the Certain World: The Life and Science of Max Born* by Nancy Thorndike Greenspan (Basic Books, 2005); Susan Quinn's *Marie Curie: A Life* (Simon & Schuster, 1995); and *Lise Meitner: A Life in Physics* by Ruth Lewin Sime, (University of California Press, 1996). The life of Nobel Laureate Enrico Fermi is described in *Enrico Fermi, Physicist* by Nobel Laureate Emilio Segre (University of Chicago Press, 1995); *Fermi Remembered*, edited by Nobel Laureate James Cronin; and *Atoms in the Family: My Life with Enrico Fermi* by his wife Laura Fermi (University of Chicago Press, 1995). I have not yet read the forthcoming *The Many Worlds of Hugh Everett III: Multiple Universes, Mutual Assured Destruction, and the Meltdown of a Nuclear Family* by Peter Byrne (Oxford University Press, 2010) but I suspect that it will be an enlightening read on at least some parallel Earths you may find yourself.

Those who would like some pictures mixed in with their words can find several excellent graphic novels that cover similar topics as those mentioned above. In particular G. T. Labs' *Suspended in Language* by Jim Ottaviani and Leland Purvis highlights the life of Niels Bohr; *Fallout*, by Ottaviani, Janine Johnstone, Steve Lieber, Vince Locke, Bernie Mireault and Jeff Parker discusses J. Robert Oppenheimer and Leo Szilard and the development of the atomic bomb; and Ottaviani's and collaborators' *Two-Fisted Science* covers, among others, Richard Feynman, Bohr, and Werner Heisenberg, while graphic discussions of quantum theory can be found in *Introducing Quantum Theory: A Graphic Guide to Science's Most Puzzling Discovery*, J. P. McEvoy and Oscar Zarate (Totem Books, 1996).

Many books have tackled the challenging task of explaining quantum mechanics while forgoing mathematics. I especially recommend *The Atom and Its Nucleus*, George Gamow (Prentice Hall, 1961); *The Story of Quantum Mechanics*, Victor Guillemin (Charles Scribner's Sons, 1968); *The Quantum World: Quantum Physics for Everyone*, Kenneth W. Ford (Harvard University Press, 2004); *The Strange Story of the Quantum*, second edition, Banesh Hoffman (Dover, 1959); Tony Hey and Patrick Walters' *The New Quantum*

Universe in a revised edition (Cambridge University Press, 2003); and David Lindley's *Uncertainty: Einstein, Heisenberg, Bohr, and the Struggle for the Soul of Science* (Anchor, 2008). Discussions of quantum entanglement and its connection to attempts to construct a quantum computer include *A Shortcut Through Time: The Path to the Quantum Computer*, George Johnson (Vintage Books, 2003); *Teleportation: The Impossible Leap*, David Darling (John Wiley and Sons, 2005); *The God Effect: Quantum Entanglement, Science's Strangest Phenomenon*, Brian Clegg (St. Martin's Griffin, 2006); *Einstein, Bohr and the Quantum Dilemma: From Quantum Theory to Quantum Information*, second edition by Andrew Whitaker (Cambridge University Press, 2006); *Entanglement* by Amir Aczel (Plume, 2003); and *The Age of Entanglement: When Quantum Physics was Reborn* by Louisa Gilder (Vintage, 2009).

Two excellent reviews of the development of solid-state physics are *Crystal Fire: The Birth of the Information Age*, Michael Riordan and Lillian Hoddeson (W.W. Norton and Co., 1997); and *The Chip: How Two Americans Invented the Microchip and Launched a Revolution*, T. R. Reid (Random House, 2001). Lillian Hoddeson's biography *True Genius: The Life and Science of John Bardeen* and Joel N. Shurkin's *Broken Genius: The Rise and Fall of William Shockley, Creator of the Electronic Age* (Palgraave MacMillan, 2006) also provide a great deal of background on the growth of this field, from the perspective of two of its founding fathers (Shockley and Bardeen were co-developers of the transistor, and Bardeen also co-discovered a microscopic theory of superconductivity, earning him his *second* Nobel Prize in Physics). The technological applications of solid-state physics are described in *Computer: History of the Information Machine*, second edition, by Martin Campbell-Kelly and William Aspray (Westview Press, 2004); *A History of Modern Computing*, second edition, by Paul E. Ceruzzi (MIT Press, 2003); *Computers: The Life Story of a Technology*, Eric G. Swedin and David L. Ferro (Johns Hopkins University Press, 2007); *Lasers and Holography: An Introduction to Coherent Optics*, second edition, Winston E. Kock, (Dover, 1981); *How the Laser Happened: Adventures of a Scientist*, Charles W. Townes (Oxford University Press, 2002); and *They All Laughed: From Light Bulbs to Lasers: The Fascinating Stories Behind the Great Inventions that Have Changed Our Lives*, Ira Flatow (Harper Perennial, 1992).

Those wishing to compare and contrast the predictions of science fiction with the reality of science may enjoy *The Science in Science Fiction: 83 SF Predictions That Became Scientific Reality*, Robert W. Bly (BenBella Books, 2005); *Different Engines: How Science Drives Fiction and Fiction Drives Science*, Mark L. Brake and Neil Hook (Macmillan, 2008); and *Follies of Science: 20th Century Visions of Our Fantastic Future*, Eric Dregni and Jonathan Dregni (Speck Press, 2006).

Background information on the history of the pulp magazines can be found in *Cheap Thrills: The Amazing! Thrilling! Astonishing! History of Pulp Fiction*, Ron Goulart (Hermes Press, 2007); *Pulpwood Days Volume One: Editors You Want to Know*, edited by John Locke (Off-Trail Publications, Volume 2007); *Alternate Worlds: The Illustrated History of Science Fiction*, James Gunn (Prentice-Hall, 1975); *Science Fiction of the 20th Century: An Illustrated History*, Frank M. Robinson (Collectors Press, 1999); *The Classic Era of American Pulp Magazines* by Peter Haining (Prion Books, 2000); *Pulp Culture: The Art of Fiction Magazines*, Frank M. Robinson and Lawrence Davidson (Collectors Press, 1998); and *The Great Pulp Heroes* by Don Hutchinson (Book Republic Press, 2007).

Those who would judge these pulps by their cover will find many excellent collections of the enduring artwork that promoted these disposable fantasies, including *Worlds of Tomorrow: The Amazing Universe of Science Fiction Art*, Forrest J. Ackerman with Brad Linaweaver (Collectors Press, 2004); *Sci-Fi Art: A Graphic History*, Steve Holand (Collins Design, 2009); *Pulp Art: Original Cover Paintings for the Great American Pulp Magazines*, Robert Lesser (Gramercy Books, 1997); *From the Pen of Paul: The Fantastic Images of Frank R. Paul*, edited by Stephen D. Korshak (Shasta-Phoenix, 2009); *Out of Time: Designs for the Twentieth Century*, Norman Brosterman (Harry N. Abrams, 2000); and *Fantastic Science-Fiction Art 1926-1954*, edited by Lester Del Ray (Ballantine Books, 1975).

Readers interested in more information about Doc Savage and his merry band of adventurers will enjoy Philip Jos Farmer's "biography" of the great man, *Doc Savage: His Apocalyptic Life* (Doubleday, 1973); a summary of the plot of each adventure is provided in *A History of the Doc Savage Adventures*, Robert Michael "Bobb" Cotter

(McFarland and Company, 2009); and Doc Savage's creator is profiled in *Lester Dent: The Man, His Craft and His Market*, by M. Martin McCarey-Laird (Hidalgo Pub. Co., 1994) and *Bigger Than Life: The Creator of Doc Savage*, Marilyn Cannaday (Bowling Green State University Popular Press, 1990). Those interested in the secrets of the Shadow (such as his *true* identity—and no, its *not* Lamont Cranston) can consult *The Shadow Scrapbook*, by Walter B. Gibson (who wrote 284 of the 325 *Shadow* pulp novels, including the first 112) (Harcourt Brace Jovanovich, 1979); *Gangland's Doom: The Shadow of the Pulps* by Frank Eisgruber Jr. (CreateSpace, 2007); *Chronology of Shadows: A Timeline of The Shadow's Exploits* by Rick Lai (CreateSpace, 2007); and *Pulp Heroes of the Thirties*, edited by James Van Hise (Midnight Graffiti, 1994).

This is a golden age for fans of Golden Age pulps, comic strips, and comic books. There are many publishers who are reprinting, often in high-resolution, large-format hardcovers, comic strips from the 1920s and 1930s, featuring the first appearances of *Dick Tracy*, *Little Orphan Annie*, Maggie and Jiggs in *Bringing Up Father*, Popeye in *Thimble Theater*, and Walt and Skeezix in *Gasoline Alley*. Several volumes of Phil Nowlan's and Dick Calkin's *Buck Rogers in the 25th Century* (Hermes Press) and Alex Raymond's *Flash Gordon* (Checker Press) are available. There are also hardcover reprints of the Gold Key comics, with at least four volumes of *Dr. Solar—Man of the Atom* by Paul S. Newman and Matt Murphy (Dark Horse Books) in print. A string of issues of DC Comics' *Strange Adventures* from 1955 to 1956 has been reprinted in black and white in an inexpensive *Showcase Presents* volume (DC Comics, 2008). Several volumes of the Marvel Comics *Tales to Astonish* and *Tales of Suspense* from this time period are also available, in *Marvel Masterworks Atlas Era* hardcovers (Marvel Publishing). The pulps themselves are also returning to print, and Sanctum Productions/Nostalgia Ventures every month is publishing classic Shadow and Doc Savage adventures from the 1930s and 1940s, often with the original interior and cover artwork reproduced. Some of the above, along with copies of *Amazing Stories* from the 1920s and 1930s, are available as e-books. We can now, in the present, download and read on our electronic book readers stories from the past, predicting what life would be like in the world of tomorrow. This is the future no one saw coming!

Note: Page numbers in italics indicate photos and illustrations.